Creating Value-Added Services and Applications for Converged Communications Networks

Creating Value-Added Services and Applications for Converged Communications Networks

Johan Zuidweg

ARTECH
HOUSE

BOSTON | LONDON
artechhouse.com

Library of Congress Cataloging-in-Publication Data
A catalog record for this book is available from the U.S. Library of Congress

British Library Cataloguing in Publication Data
A catalog record for this book is available from the British Library.

ISBN-13: 978-1-60807-786-1

Cover design by John Gomes

© 2015 Artech House
685 Canton St.
Norwood, MA

10 9 8 7 6 5 4 3 2 1

To Ana, Daniel, and Hugo

Contents

Acknowledgments

I would not have been able to write this book without the support of professionals, colleagues, and friends. First and foremost, I'd like to thank the reviewer of my book, Dr. Juha Korhonen of ETSI, who contributed immensely to improving the quality of the manuscript. I also want to thank Aileen Storry from Artech House for her management of the acquisition and review process.

My special thanks go out to Dr. Miquel Oliver, who leads the Network Technologies and Strategies (NeTS) research group of the Universitat Pompeu Fabra (UPF), for his confidence in me and his continued support. I'm grateful to José Pérez, the CEO of Tecsidel S.A. in Spain, for letting me bounce my ideas off him and providing me with priceless feedback.

I draw a lot of inspiration from my friend Dr. Edwin Paalvast, who runs Cisco's service provider business in a significant part of the world. I'm endebted to Musa Unmehopa (M.Sc. and MBA) from Philips for teaching me about leadership and standards strategy. I thank my friend and former colleague Dr. Hugo Velthuijsen of the Hanze University of Applied Sciences in Groningen, The Netherlands, for enlightening me with his remarkable strategic insight and broad field of interest. Our conversations often go beyond tech talk and touch upon culture, philosphy, and life itself.

An author's family tends to pay the highest price for a new publication, and it has been no different in my case. I owe apologies to my sons, Daniel and Hugo, for the times when I should have been playing football, ping pong, or chess with them and instead sat staring at a computer screen. I don't know how to compensate Ana for the moments together we missed because I had to deliver the next chapter.

Preface

In the fall of 2013, Artech House asked if I'd be willing to publish a second edition of my 2002 book, *Next Generation Intelligent Networks*. It took me no time to say yes, but I immediately realized that the book needed more than a few revisions. So much has happened in the past decade that I had to rewrite it completely.

In the decade following the publication of *Next Generation Intelligent Networks*, IP networks conquered the world and now routinely carry voice telephony, too. In addition, 3G networks have brought the Internet to mobile devices, and more recently LTE enables serious data speeds. Smart phones and application stores have blurred the lines between information and communications technology, and over the top (OTT) services have uprooted the network operators' business. Cloud computing is changing the way enterprises and consumers use software and store their data. Machine-to-machine (M2M) communications are pointing to a world where everything is connected. Technologies such as the session initiation protocol (SIP) and the internet multimedia subsystem (IMS) have matured, while others such as PINT, SPIRITS, and WAP have faded (I won't spell out these acronyms here).

These changes in the communications landscape made the prospect of writing a new book all the more relevant and appealing, so I set out to write what I can best describe as a sequel to 2002's *Next Generation Intelligent Networks*. The book you presently have in your hands aims to capture state-of-the-art technologies that enable service providers to create and deploy value-added communications services.

SIP and IMS play prominent roles here because they leverage IP networks. Java has become the most extensively used programming language in communications and also receives considerable attention in this book. At the center of this book are application programming interfaces (APIs) for networks and devices, which have become key enablers for third-party applications and have followed the information technology (IT) industry's trend toward web services based on representational state transfer (REST).

I reference several excellent publications on SIP, IMS, and Java in this book, but each tends to concentrate on a specific topic and in some cases provides either too much or too little technical detail. I know of few books that address the creation of services for communications networks from a more comprehensive viewpoint. I missed a book that gives a complete but concise overview of enabling technologies with sufficient technical detail to understand their technical specifications, yet without losing sight of the overall picture. I hope I succeeded in creating the book that fills this gap.

I've written this book for a broad audience including managers, developers, and academics. I've tried to structure it in such a way that managers can get an understanding of the service enablers without having to wade waist deep through technical details. Developers, on the other hand, can choose to dive in deep and search for specific interfaces of their interest. Academics and university professors can use the material to teach their students about service creation and deployment in communications networks.

In this book, I explicitly avoid spelling out technical specifications to parameter or byte level. This would clutter the book and make it more difficult to see the forest through the trees, and it's futile to reproduce specifications you can find on the portals of standards organizations. Instead, this book will help you understand the concepts behind the specifications and determine what to read and how to interpret it. Where needed, I'll point you to the original specifications so you can download them from the web should you need more technical detail.

Wherever possible, I have tried to refer to international rather than regional standards. Since I live in Europe, you may at times notice some bias toward groups like the Third Generation Partnership Project (3GPP) and the European Telecommunications Standards Institute (ETSI) rather than, for example, the American National Standards Institute (ANSI), the Telecommunications Technology Association (TTA), or the China Communications Standards Association (CCSA). However, I have tried to avoid prejudice wherever possible and focus on concepts that are common to all standards.

Things are still very much in flux, and it will be difficult to predict what will happen in the next 10 years. But it is almost certain that mobile broadband will continue driving the future, and the Internet will fuel many innovations to come. I'm confident that the technologies I've described in this book will remain relevant for a decade or more.

I much appreciate any comments you, the reader, have on this publication. You can communicate them to me by sending mail to johan.zuidweg@ ieee.org or johan.zuidweg@gmail.com.

1

Introduction

This book is about information technology in communications, "IT" in "C." More specifically, it's about building applications and services over communications networks and managing them. If you're reading this book now, you're most probably involved in this business in some way. This introduction will give you a detailed view of the scope and contents of this book, whom it is intended for, and what you can expect from it. Let's begin by setting the scene.

1.1   How We Got Here

Through humanity's history, people have used fire, mirrors, flags, bull roarers, pigeons, lighthouses, letters, radios, telegraphs, telephones, and the Internet to bridge the distance between them. Many consider the invention of the telegraph the beginning of modern telecommunications.

The idea of the telegraph goes as far back as the mid-1700s, and in 1838 Samuel Morse[1] made it practical by inventing a way of sending information over a single line as a sequence of electric pulses of varying length.

---

[1] Interestingly, American-born Samuel Finley Breese Morse (1791–1872) dedicated the first half of his life to painting rather than invention.

He represented each character in the alphabet with a unique combination of long and short pulses (dots and dashes), a scheme that we still know today as Morse code.

The U.S. government immediately seized the importance of this invention and from then on things moved fast. In 1843, the first public telegraph connection was inaugurated between Washington and Baltimore, and in 1866 the first transatlantic telegraph cable was in place. By the end of the 1860s, news traveled around the world faster than ever before.

Even in the early days of the telegraph, people thought of ways using telegraph lines for voice communications. Graham Bell invented a mechanism to transform sound into an electric signal and demonstrated the first telephone in 1876. The Bell Telephone Company he helped create in 1877 became the world's most influential telecommunications company for more than a century. By 1899 it adopted the name AT&T.

In the decades that followed, manual switchboards gave way to automatic, electromagnetic exchanges, and telephones changed manual ringers to rotary dials. The telephone network grew spectacularly in the number of land lines, exchanges, and subscribers, but telephony as a service changed very little for almost 100 years.

In the 1980s, however, advances in computing began causing important transitions in telecommunications. Computers enabled digital signaling, which made it possible to provide value-added supplementary services such as call forwarding, call blocking, call waiting, and three-way calling. The intelligent networks (IN) standard published by the International Telecommunication Union (ITU-T)[2] in the early 1990s describe the architecture to provide such valu-added services without standardizing the services themselves.

Progress in computing also enabled digital cellular networks such as GSM, which brought about another breakthrough in telecommunications. The proliferation of small and inexpensive computers, and most prominently the personal computer (PC), fueled the explosive growth of the Internet and the World Wide Web (WWW) throughout the 1990s and 2000s.

Until the turn of the millennium, the Internet and the telephone still represented mostly separate worlds, exemplified by the terms *net head* (an Internet proponent or specialist) versus *Bell head* (somebody from the world of circuit

---

[2] You'll have noticed that there is an extra "-T" in ITU-T. ITU-T refers to the Telecommunication Standards sector of the International Telecommunication Union. If this sounds like a pleonasm, bear in mind that the ITU also has a radiocommunications sector, abbreviated as ITU-R. Whereas the ITU-T standardizes protocols and IT aspects of telecommunications, the ITU-R allocates radio spectrum and publishes recommendations to make optimal and interference-free use of that spectrum.

switched telephone networks). Voice over IP (VoIP) did not yet compete in earnest with the good old telephone, but new Internet-based applications such as instant messaging (IM) and chat heralded a new era in communications.

Things evolved dramatically throughout the 2000s when digital subscriber lines and fiber brought broadband Internet to our homes and 3G mobile networks connected mobile devices to the Internet on a widescale speed. Now there was enough coverage and bandwidth to use the Internet for almost anything, including voice communications and broadcasting radio and television.

Reaching a critical mass of connected individuals, the Internet did not take long to produce a host of new applications such as social networking, media sharing, conferencing, and online gaming, blending real-time communications with data. Ubiquitous connectivity also created a new breed of devices such as the smart phone, the tablet, and wearables. Lucky entrepreneurs who are in the right place at the right time make billions of dollars with Internet applications, some of them of striking simplicity.

With shrinking revenues from providing communications services and expensive infrastructures to maintain, network operators find themselves at the losing end of the game. Regulation in many jurisdictions does little to help them by increasing competition, putting pressure on prices and obliging service providers to give equal treatment to all Internet traffic, including high-bandwidth traffic from third parties competing with over the top (OTT) communications services.

But operators also take advantage of the opportunities created by the Internet. The internet multimedia subsystem (IMS) enables them to provide carrier-grade voice and multimedia communications over IP networks, using the session initiation protocol (SIP) for signaling. Through application programming interfaces (APIs) accessible as web services, operators can provide business-to-business (B2B) services to third party service providers and recover part of the revenues they lose to OTT players. Machine to machine communications and cloud computing provide new opportunities, too.

This book is about the new technologies and standards that allow operators and service providers to create and deploy value-added services in a changing world increasingly dominated by packet switched networks using the internet protocol (IP).

1.2 Who'll Benefit from This Book?

If you're reading this book, you're presumably engaged in communications services and networks, maybe as a developer or manager. Or perhaps you're a student who needs to know about communications technologies. In all these cases, this book has useful content to offer you.

If you manage telecommunications business or technology, you will have a complete view of legacy and state-of-the-art technologies available to deploy and manage value-added communications services. Each chapter provides an introduction to basic technology and is mostly self contained. The way this book is written will let you skip technical details you don't need to know without losing the overall picture. It will enable you to judge what technology serves your business where and how.

If you develop applications or services, this book will provide you with an extensive overview of the APIs available at several levels of abstraction. You'll find descriptions of lower-level Java APIs that allow for direct manipulation of SIP messages, as well as web services that allow you to use network services such as call control, messaging, and location. Navigating standards organizations to find a particular specification (and the right version!) is often a daunting task. This book will direct you to the sources where you can download the specifications you will need in your work.

If you teach telematics or telecommunications courses, you can use this as a text book to explain common channel signaling system number 7 (SS7) and SIP signaling, IMS, Java for communications applications, network and device side APIs, as well as basic network management and charging concepts. Each chapter contains an introduction into the basic technology needed to understand the more advanced content.

After reading this book you'll have the knowledge to explain to others the meaning of IN, CAMEL, SIP, IMS, RCS, MMTel, JSIP, SIP Servlets, JAIN SLEE, REST, OSA, Parlay X, oneAPI, OMA RESTful network APIs, OSS, BSS, CDR, IPDR, Frameworx, MTOSI and many more acronyms and buzzwords (for now, we won't explain them here). You will know:

- How intelligent networks (IN) and customized applications for mobile network enhanced logic (CAMEL) enable the creation and deployment of value-added services in circuit switched fixed and mobile networks;
- How a prepaid calling with CAMEL works;
- What purpose and structure SIP has, and how IMS uses this protocol;
- How to create and deploy value-added services in IMS;
- What interfaces Java provides for programming communications applications based on SIM and IMS;
- What standard APIs exist for networks and devices to build value-added communications applications and services with;

- Which functions an operations support system (OSS) and business support system (BSS) have, and which standards they build on;
- How OSS charge services and network resources, and which standards exist for charging;
- What threats and opportunities machine-to-machine (M2M) communications and cloud computing bring to network operators.

1.3 Prerequisites and Technical Depth

This book aims at explaining technologies in technical detail, but without losing the context and overall view. Its chapters treat architectures, interfaces, protocols, and standards in depth, but do not specify protocols to the byte level, do not describe interfaces down to every parameter, and do not provide executable Java code.

The text explicitly avoids reproducing specifications published as standards. Standards organizations' portals and search engines have made it so easy to find and download specifications that there is no point in rehashing information you can find on the Internet. Instead, this book has the purpose to help you identify the relevant standards, appreciate how to read them, and know where to find them. It tries to hit the sweet spot between a readable synopsis and full technical detail.

For example, this book concentrates on IMS functions that provide SIP signaling functions critical to the implementation of value-added services, but does not enumerate all IMS functions engaged in media transfer and control. It dedicates a good deal of attention to APIs and gives an extensive overview of the scope of their interfaces, but does not list them down to operation and parameter level. The chapter on Java explains relevant language constructs and contains example code snippets, but stops short of listing code you can actually compile. Excellent books exist on IMS [1, 2] and Java [3, 4] that will help you, should you need to read up on these technologies.

Although each chapter goes into considerable technical detail, this book assumes only basic knowledge of computing and communications technologies. Each chapter contains an introduction that explains the fundamental technology needed, in plain English and avoiding jargon wherever possible. You do not need expertise on signaling, IP, mobile networks, Java, SIP, IMS, web services, or network management to read this book and grasp the concepts explained.

If you already know about these technologies, you can skip the introductory sections. Some experience with computering, programming languages, object-oriented design principles, and IP will come in handy but is not strictly required to master the material in this book.

1.4   The Contents of This Book

This section gives a more detailed overview of the contents of this book, and the information you'll find in each chapter. Section 1.6 suggests some specific paths through the sections of this book, depending on your particular area of interest.

1.4.1   Creating Value-Added Services in Circuit Switched Networks (Chapter 2)

Chapter 2 discusses standards to create value-added services in fixed and mobile telephone networks. It starts with a brief tutorial on circuit switched telephony and digital signaling with SS7. Chapter 2 discusses in detail two standards: the IN specifications for fixed networks and CAMEL specifications for mobile networks. Chapter 2 concludes with an elaborate example that shows how CAMEL enables prepaid mobile calling.

Given the advance of packet switched networks replacing circuit switched networks, some may argue that Chapter 2 addresses legacy networks that have seen their best days. Yet many operators around the world still support circuit switched telephony, and IN and CAMEL will still be there for some time to come.

In a nutshell, Chapter 2 covers:

- An introduction to circuit switched telephony networks and SS7 signaling;
- The IN conceptual model and service plane, global functional plane, distributed functional plane, and physical plane;
- An introduction to mobile networks and their specific signaling procedures;
- CAMEL as an adapted version of IN for mobile networks;
- Implementation of a prepaid mobile calling service with CAMEL.

1.4.2   Value-Added Communications Services in Packet Switched Networks (Chapter 3)

The third chapter of this book explores standards for delivering value-added voice and data services over packet networks. The first part of this chapter reviews the basics of IP addressing and routing, and explains how cellular mobile networks provide IP access. The core of this chapter describes IMS and the protocols that enable it, in particular SIP.

Chapter 3 puts special emphasis on the role of the application server (AS) in IMS, which hosts value-added applications. It also reviews two IMS-based standard services, the rich communications suite (RCS) and multimedia telephony (MMTel). In a nutshell, Chapter 3 covers:

- A summary of IP addressing and routing principles, and IP protocols;
- How cellular mobile networks support IP;
- Protocols for providing voice and multimedia communications over IP, with emphasis on SIP;
- Building services with IMS components and procedures;
- IMS-based standard services: MMTel and RCS.

1.4.3 Programming Communications Applications and Services with Java (Chapter 4)

Chapter 4 is dedicated entirely to Java, the standard programming language widely used to implement applications and services. After reviewing the basics of object-oriented programming and Java essentials such as the virtual machine, byte code, and the Java API, this chapter examines in detail a number of Java facilities to manage SIP-based communications and create value-added IMS applications.

In a nutshell, Chapter 4 covers:

- An introduction to object-oriented programming and Java;
- The Java SIP (JSIP) API to handle SIP signaling in a Java program;
- Managing SIP signaling with SIP servlets;
- The Java APIs for integrated networks (JAIN) service logic execution environment (SLEE) for carrier grade Java applications;
- The Java IMS API.

1.4.4 Application Programming Interfaces for Distributed Service Control (Chapter 5)

In the late 1990s, both technological and regulatory factors drove manufacturers and service providers to begin introducing distributed object-oriented design into their systems. Standards organizations followed suit by specifying APIs that enable service providers to expose their network functionality to third parties. Third parties can use these APIs to create value-added applications for IMS and for circuit switched networks (IN and CAMEL).

Chapter 5 provides an extensive overview of the network and device APIs specified by the Third Generation Partnership Project (3GPP), the Open Mobile Alliance (OMA), and the GSM Association (GSMA). It first reviews key enabling technologies such as the extended markup language (XML), simple object access protocol (SOAP), web services definition language (WSDL), and representational state transfer (REST). The main part of this chapter lists the main features of each API specification and explains how each API enables prepaid calling, allowing you to compare them through an example.

In a nutshell, Chapter 5 covers:

- An introduction to distributed object-oriented systems and XML;
- Web services: SOAP, WSDL, and REST;
- Network side APIs: Parlay, OSA, Parlay X, OMA REST Network APIs, and GSMA OneAPI;
- Examples of a prepaid calling service implemented with OSA, Parlay X, and OneAPI;
- Device side APIs: SIM application toolkit (SAT), mobile execution environment (MExE), device management (DM) client API, and open connection manager API (openCMAPI).

1.4.5 Support Systems (Chapter 6)

The technologies and standards discussed in Chapters 2 through 5 have no purpose unless deployed in an environment where the service provider can manage its subscribers, its services, and the network resources they rely on. Most importantly, the service provider must be able to charge for services and resources used, and bill them to the subscribers who used them.

This chapter provides insight into the systems that support an operator's or service provider's operations and business: the OSS and BSS. Particular attention goes to charging and billing aspects, as well as network management techniques. Although these subjects merit a complete book of their own, Chapter 6 provides a concise outline of the main concepts and standards that enables basic understanding of the network and service management domain. In a nutshell this chapter covers:

- The functions of OSS and BSS;
- Network management concepts and specifications used in OSS and BSS;

- Charging and billing procedures and standards;
- Service delivery platforms (SDPs).

1.4.6 Things to Come (Chapter 7)

The last chapter of this book looks forward and identifies technological trends that have the potential to shape telecommunications of the future, both in terms of technology and business. OTT applications, cloud computing, M2M communications, and the Internet of things (IoT) represent evident areas of technological innovation and disruption. Chapter 7 takes a closer look at these subjects and tries to identify the threats and opportunities for the network operator and service provider. In a nutshell, this chapter covers:

- Machine to machine communications and the oneM2M Partnership Project;
- Cloud computing and virtualization;
- Network centric versus over the top services;
- The future of the network operator in a changing environment.

1.5 Timing

Unlike a blog or website, a book can only provide a snapshot of whatever has relevance at the time of writing. Information and communications technologies evolve with notorious speed, so this book will undoubtedly lag behind the latest advancements by the time you have it in your hands.

The text in this book draws heavily on standards. Standards tend to trail behind state-of-the-art technology, but they also have more stability because they are the fruit of judicious negotiations between stakeholders. While standards may not always describe the technologically best or most advanced solution, they do often provide sustainable compromises for the technological tradeoffs and conflicts of interest that exist in the industry. The standards-oriented approach also avoids that this book biases a particular proprietary product or solution.

Standards do undergo changes, too, and not all standards gain wide acceptance in industry. Some may argue that the IN and CAMEL standards are on the way to obsolescence, but many operators continue deploying IN and CAMEL and the ITU-T still sustains the specifications.

The technologies in this book find themselves at different stages of maturity and industry acceptance. The Gartner hype cycle illustrated in Figure 1.1

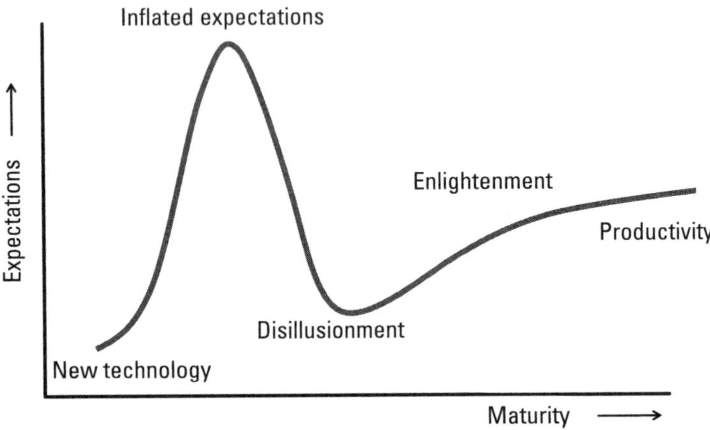

Figure 1.1   Gartner hype cycle.

provides a popular model for the level of visibility a technology has in industry in function of its maturity. The Gartner model describes how new technologies tend to go through a hype, followed by a period of disillusionment when the industry becomes aware of the inflated expectations. Eventually, some technologies (not all) will mature to the point where the industry embraces the technology for what it can and cannot do, and moves to sustainable deployment.

One could argue that IMS and many of the network APIs described in this book have passed their hype and traverse their trough of disillusionment, while Java and SIP have moved on to acceptance and consolidation. M2M communications and cloud computing appear to be gliding down the slope of their hype, too, and will have to go through their redemption cycle before we can consider them mainstream technologies. But all the standards and technologies treated in this book have a solid chance to eventually progress to mature deployment and exploitation.

1.6   Reading This Book

You can read this book in different ways, depending on your subject of interest and on what you know already. The chapters of this book have a logical sequence. You may consider reading all chapters in order if you need to get the complete picture of the technologies available to create value-added services and applications in communications networks.

Each chapter covers a designated subject and reviews the basic technology needed for its understanding before going into details. If you are interested in a particular subject, for example, CAMEL, IMS, or Java APIs, you can go directly to the chapter in question.

It is also possible to read combinations of chapters that highlight a particular approach to value-added service creation. Possible tracks through the material in this book are

- *Intelligent networks track:* if you need information on IN and CAMEL, then read Chapter 2. The APIs discussed in Sections 5.3 and 5.4 in Chapter 5 also have relevance because they also apply to IN and CAMEL, so you may consider taking those into account, too.
- *IMS track:* if you're primarily interested in building applications and services for IMS, then go directly to Chapter 3. If you already know the basics of Internet protocols, you can skip Section 3.1.
- *SIP track:* if you need to learn about SIP signaling you should read a combination of sections in Chapters 3 and 4, in particular Sections 3.3, 4.3, 4.4, and 4.5. If you're new to Java, read Sections 4.1 and 4.2, too.
- *Java track:* Chapter 4 is dedicated exclusively to Java programming for SIP and IMS. It contains an introduction to Java concepts in case you are not familiar with them.[3]
- *API track:* Chapter 5 extensively discusses network- and device-based APIs that enable the creation of value-added services. If you know the basics of distributed object-oriented programming, you can skip Section 5.1. If you are familiar with web services technology (XML, SOAP, WSDL, REST), skip Section 5.2, too, and go directly to the discussion of specific APIs starting in Section 5.3. If you're interested only in more recent state-of-the-art APIs, you can limit your attention to Sections 5.4 and 5.5. In case you need only information about APIs in the device, go directly to Section 5.6.

Although specifically bound to the Java environment, JAIN SLEE also provides network APIs, and you should consider including Section 4.5 in your reading. In addition, the Java IMS API in Section 4.6 represents a Java specific device side API.

---

[3] A word of caution: you learn programming mostly by doing. Although Chapter 4 of this book tries to explain Java programming from the ground up, it will not be easy to grasp the feel of Java programming if you have never had your hands on a keyboard. You certainly don't need to be an ace programmer to understand Chapter 4, but having had a primer on Java and having experienced programming your own "Hello World" code will certainly help in digesting this chapter.

- *Management and charging track:* although one of the shorter chapters of this book, Chapter 6 provides a concise overview of network management concepts, with particular attention on charging.

References

[1]   Camarillo, G., and M. A. García Martín, *The 3G IP Multimedia Subsystem (IMS): Merging the Internet and the Cellular Worlds*, Chichester, UK: John Wiley & Sons, 2008.

[2]   Martinez Perea, R., *Internet Multimedia Communications Using SIP: A Modern Approach Including Java® Practice*, Burlington, MA: Morgan Kaufmann, 2008.

[3]   Schildt, H., *Java, A Beginner's Guide*, New York: McGraw-Hill, 2011.

[4]   Bloch, J., *Effective Java*, Boston, MA: Addison-Wesley, 2008.

2

Value-Added Services in Circuit Switched Networks

Today's generation has many ways to communicate: by telephone, cell phone, short message, e-mail, chat, and video, to name just a few. But it's not more than 25 years ago that the telephone was the only means of telecommunication for most people. Public switched telephony has served humanity for more than a century, from the late 1800s to the present day.

The invention of the telephone by Alexander Graham Bell and his assistant Thomas Watson came only a few decades after the invention of the telegraph. It was in many ways a logical next step. Bell figured that if one worked out how to convert the human voice into an electrical signal, the signal could be sent over the telegraph wires that were already deployed, making communications more efficient and more immersive. He demonstrated his invention in 1876, less than 30 years after Samuel Morse invented the telegraph.

The telephone was revolutionary in that it almost instantly became part of the household. As early as the beginning of the 1900s important citizens already had a telephone in their residence. By the 1950s the majority of American and European households had a telephone.

The Internet today challenges the public switched telephone network by offering alternative ways of communicating, including voice, but switched telephone networks are still widely used. For this reason, and because Internet-based communications do take inspiration from telephony (the notion of a call never changes), we have to take switched telephony seriously even today.

This chapter takes a detailed look at the technology behind telephone networks, fixed and mobile, and how they are used to provide value-added communications services beyond simple voice calls. This also provides the essential background for later chapters that delve more deeply into Internet-based and mobile communications.

2.1   The Public Switched Telephone Network

What characterizes the public switched telephone network is that it enables voice communications by creating a connection between the caller and callee. In early telephone networks, the connection was an electrical circuit carrying the analog voice signal. In the digital age circuits became virtual and digital, but they're still circuits, and this is why we say that telephone networks are circuit switched.[1]

Until the end of the nineteenth century, telephone networks were manually operated. All telephones lines terminated in a central office, where operators manually established circuits by plugging lines into the appropriate jack. These central offices were operated by many different companies and offered only local service. There were few long distance connections. Figure 2.1 illustrates the concept of the central office.

The explosive demand for telephony soon created the necessity for automated establishment of voice circuits. This led to the invention of electromechanical telephony exchanges (also called telephone switches)[2] that were used for at least half a century. An electromechanical exchange uses a system of electrical relays that respond to the electric pulses made by a telephone's rotary dial (1 = one pulse, 2 = two pulses, and so on).

The demand for telephony over larger distances introduced the need for long distance carriers, telephone companies that are able to interconnect

---

[1] At present, many telephone networks also use packet switching, as we shall see in Chapter 3 and beyond. Until the end of the 1990s, public telephone networks almost exclusively used circuit switching, and today there are still many circuit switched telephone networks in the world. The technologies discussed in this chapter apply to those circuit switched networks, while in the next chapter we'll turn to packet switched networks.

[2] Since telephone exchanges switch connections between telephones, people often call them switches. Strictly speaking, the switch is only part of an exchange and refers to the matrix that does the actual switching between circuits. An exchange has additional components such as the signaling terminations and its control module. Throughout this book we'll therefore use the term *exchange*, but we'll use the term *switch* wherever exchange would cause confusion.

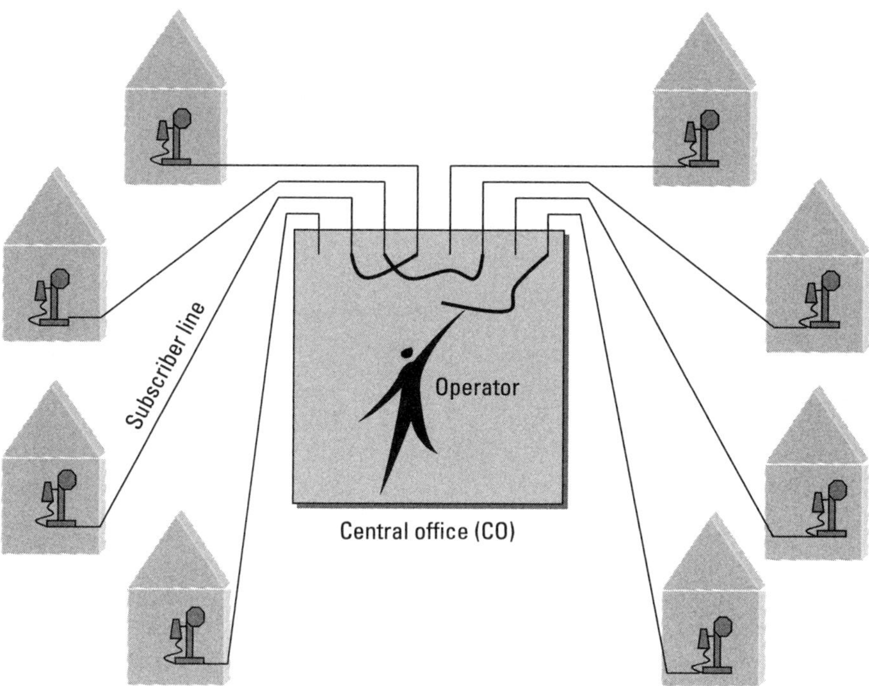

Figure 2.1   The concept of a central office.

central offices between cities, regions, and even countries. AT&T was the first long distance carrier and remained the most important one until well into the twentieth century.

To allow for communications between cities, states, regions, or even countries, central offices were connected to special interconnection exchanges. This eventually led to an interconnection hierarchy with five levels or classes, as shown in Figure 2.2. In this hierarchy, the central office is the lowest level and is called a class 5 exchange. The exchanges at all other levels only switch connections between other exchanges and have no subscribers attached to them. The lower the class number, the higher in the hierarchy it is and the bigger the exchange.

2.1.1 Digital Telephone Exchanges

By the 1960s the transistor had found its way into switching systems, giving way to a new generation of electronic exchanges. The only mechanical component in these exchanges was a matrix of tiny reed switches controlled by

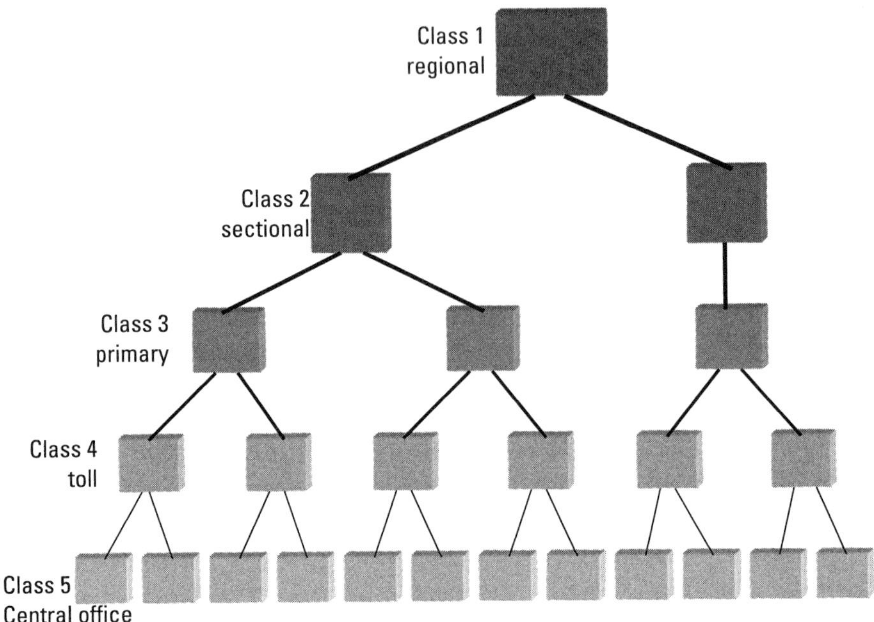

Figure 2.2   Hierarchy of telephone exchanges.

electronic circuits. These electronic exchanges also used tones, known as dual tone multifrequency (DTMF), rather than pulses to dial numbers.

The development of the computer led to a new generation of telephone exchanges in the early 1970s. These exchanges were in effect computers controlled by stored programs. In the second half of the 1970s these computerized exchanges became completely digital and didn't have any mechanical parts anymore.

Rather than physically establishing an electrical voice circuit, digital exchanges convert the voice signal in a series of numbers encoded in 1s and 0s (bits) and send these bit streams electronically from source to destination. Multiple bit streams are multiplexed into one high-speed bit stream sent on the trunk line between exchanges and demultiplexed into separate streams again at the other end. Figure 2.3 illustrates this in very simplified form.

In the more recent generations of exchanges even the subscriber line can be digital by eliminating the analog-digital conversion (see Figure 2.3). This integrated digital services network (ISDN) allows a single subscriber line to be used for telephony, fax, and data communications; however, it requires a digital network termination to be installed in the customer's premises.

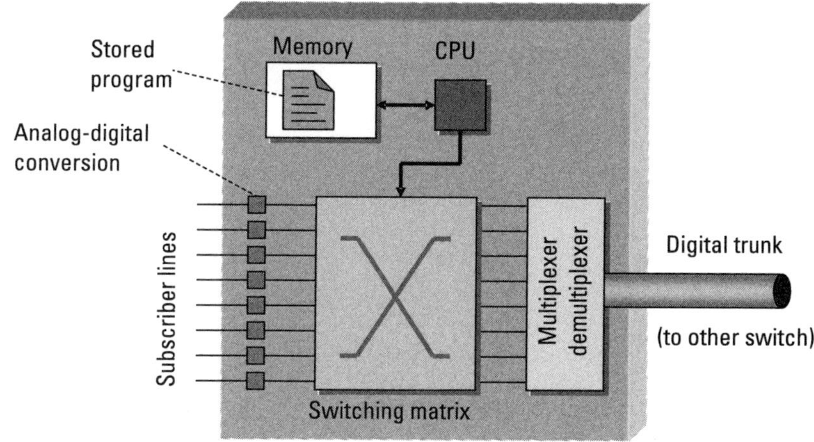

Figure 2.3　Concept of a stored program digital exchange.

Stored program digital exchanges can be programmed like any other computer. AT&T and other network operators started adding little programs to digital exchanges, allowing them to do more than just set up a voice circuit to a subscriber or another exchange. These stored programs allowed the dialed number to be processed and the call destination to be manipulated before the call is set up, enabling new services such as freephone (800 numbers, or toll-free numbers), call forwarding, and call blocking.

2.1.2　Call Signaling

The digitalization of exchanges also marked an important evolution in call signaling. Signaling is the communication necessary to set up a call from one subscriber to another. There are two kinds of signaling: user-to-network signaling (e.g., the numbers dialed to make a call) and network-to-network signaling (the messages that telephony switches need to exchange in order to realize the call).

In the early days, signaling was done on the same lines that also carried the voice circuits (in-band signaling). This meant that a telephone exchange had to set up a complete connection just to ring the destination subscriber and tear it down again if a busy signal or no reply was received. It is plain to see that a lot of capacity was wasted this way.

Engineers at AT&T therefore proposed a new system whereby the signaling was done over separate connections. They called it common channel signaling system number 7 (SS7). Since digital exchanges are now computers,

SS7 is nothing more than a very reliable computer network, built on the principles of packet data communications. SS7 is still in widespread use today, and it has had a profound impact on the evolution of the telephone network since the 1980s.

2.1.3 Common Channel Signaling System Number 7

As electronic telephone exchanges became widespread in the 1970s, the need arose for modernizing the signaling protocols. AT&T designed SS7 in 1975 using network technology that was then being used to interconnect computers with the objective to guarantee very reliable signaling links.

In 1980 ITU-T published the first SS7 standard as the so-called Yellow Book. Updates followed in 1984 and 1988, and ITU-T has maintained and updated its SS7 standard ever since. Unfortunately there is not one standard of SS7. The ANSI and ITU-T versions of SS7 are slightly different, and there are several different versions of SS7 deployed around the world. There are, however, gateways that interface between the different protocols.

A SS7 network contains three kinds of elements:

- *Signaling points* (SPs): network equipment that can send or receive signaling messages.
- *Signaling links* (SLs): links that carry signaling messages. They can be low-speed (56 Kbps or 64 Kbps) or high-speed (1,536 Kbps or 1,984 Kbps) data connections.
- *Signaling transfer points* (STPs): intermediate nodes that route signaling messages from one place to another.

Figure 2.4 shows a typical SS7 configuration for two connected networks. STPs are usually configured in pairs to ensure high reliability. The SS7 protocols ensure that when an STP or SL goes down, the messages are automatically rerouted over links that are still operational.

To ensure secure and flexible signaling connections, the SS7 protocols are organized in layers. Each protocol layer provides certain services to the layer above it and uses the services of the layer below. This results in a modular protocol stack where each layer encapsulates a well-defined set of functions.

Figure 2.5 shows the SS7 protocol "stack," which is made up of the following protocols:

- *Message transfer part 1* (MTP1): represents the physical layer consisting of hardware and firmware resources like network cards, transceivers, and cables.

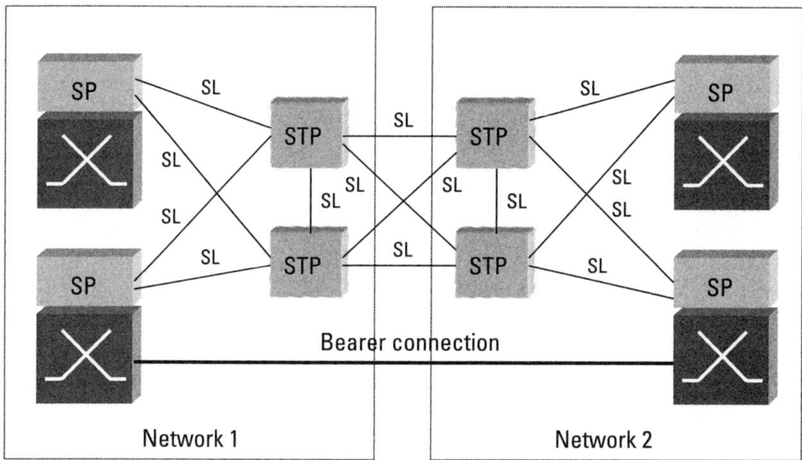

Figure 2.4 SS7 network elements.

- *Message transfer part 2* (MTP2): is responsible for the secure transmission of messages between two signaling (transfer) points.
- *Message transfer part 3* (MTP3): takes care of routing a message from one point in the SS7 network to another, passing through intermediate signaling transfer points.
- *Telephony user part* (TUP): describes the signaling messages for the setup of calls and connections in analog telephony networks. This protocol is used in older telephony networks that still use electromechanical or analog electronic exchanges.
- *ISDN user part* (ISUP): describes the signaling messages for the setup of calls and connections in digital networks. Despite the name, ISUP is used in all digital telephony networks, not only ISDN.
- *Signaling connection control part* (SCCP): sets up and manages signaling connections, using MTP3 to route messages reliably from one node to another. In fact SCCP supports both connection oriented and connectionless signaling contexts. SCCP also carries the information that STPs need to perform global title translation. This translation allows global titles (dialed numbers) to be translated to destination point codes (network addresses). This is needed for translating 800 numbers, number portability, and other services where the dialed number must be translated to a different destination address.
- *Transaction capabilities application part* (TCAP): allows signaling nodes to do transactions. Transactions are groups of actions that must be

performed as one indivisible action. They are typically used for accessing network databases (e.g., databases that map 800 numbers to their destination address). A TCAP message contains two types of information: the *transaction portion* is for starting and ending transactions, and it maintains the state of the dialog; the *component portion* carries the actual protocol queries and responses.

A TCAP message is a kind of container that can carry the signaling messages of other protocols. Some protocols that were defined are:

- *Operation, administration, and management part* (OAMP): a protocol used to manage network routing databases and to diagnose link problems.
- *Mobile application part* (MAP): the protocol for mobility management in digital cellular mobile networks such as GSM.

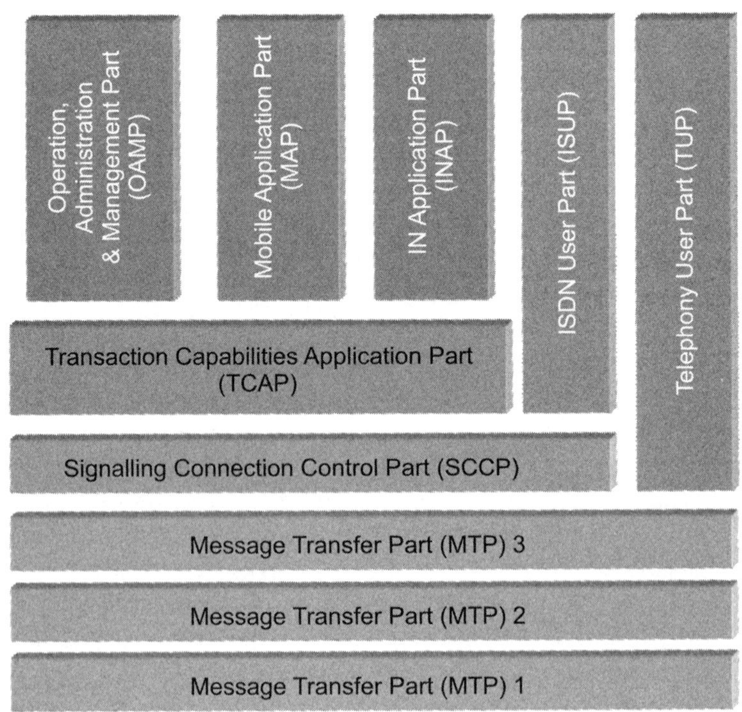

Figure 2.5   SS7 protocols.

- *IN application part* (INAP): the protocol defined for intelligent networks (IN). We will look at both MAP and INAP in more detail in this chapter.

By now you may have observed that telephone exchanges are essentially computers connected over a very reliable data network and with a dedicated purpose: to set up and tear down circuits for voice communications. And digital exchanges were being programmed to process simple services such as call forwarding, call blocking, and freephone numbers.

At first, these services were programmed in assembler-like languages, and had to be installed manually in each exchange. Installing a new service meant reprogramming the exchange and manually resolving the interactions with the service programs already running on it. The process of defining the service, specifying, implementing, testing, and finally deploying it would typically take between two and four years. The management of these services was obviously a nightmare, as even the smallest patch required going to each exchange in turn and reinstalling the new software.

Bell Communications Research Inc. (Bellcore, later Telcordia, and now part of Ericsson) came up with the idea of using the SS7 network to process and manage these services in a central node rather than in the individual exchanges. It is obviously more efficient, more secure, and less costly to deploy and manage these services in a central node than in many distributed exchanges.

Bellcore therefore proposed the addition of a service control point (SCP) to the SS7 network to offload additional service processing from individual exchanges, as illustrated in Figure 2.6.

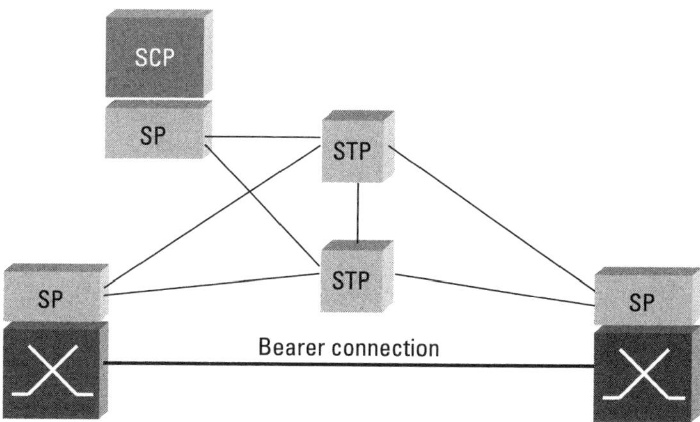

Figure 2.6 Introduction of service control points in SS7 networks.

Consider the example of call forwarding. When a subscriber makes a call, the exchange she is connected to first routes the call setup signaling message to the SCP. The SCP looks at the dialed number, and if it is to be forwarded, it substitutes this number with the new destination number. The resulting modified signaling message is then sent onward into the SS7 network to complete the call.

Bellcore named this idea of adding SCPs to the SS7 network the advanced intelligent network (AIN), and it laid the foundation for the international IN standards to follow. The next section discusses how IN was further developed and standardized.

2.2 Intelligent Network

When Telcordia successfully implemented its AIN architecture in the late 1980s, the idea caught on with the international telecommunications community and ITU-T began efforts to standardize it. The ITU-T took four years to publish the first IN standard in 1992 as the Q.1200 series, and by that time the first products were already coming to the markets. The IN standards from the ITU-T remain close to Bellcore's AIN in concept but differ on a protocol level.

The main philosophy behind the standardization process was to standardize service capabilities, not the services themselves. This is important in understanding the IN architecture. Many people talk about IN services but these in fact don't exist, at least not in terms of standards. What exists are services that can be implemented with IN, but there is no such thing as a service standardized by IN.

Although the ITU-T initiated IN standardization, it is not the only standardization organization that has shaped IN standards. The European Telecommunications Standardization Institute (ETSI) has made several important contributions to IN standardization by defining core-INAP and CAMEL, discussed in Section 2.8. The T1S1 group of the American National Standards Institute (ANSI) has contributed to the American standardization of IN.

Other standards organizations and industry groups have built upon IN standards to create other specifications and standards. Examples are the Third Generation Partnership Program (3GPP), the Java Applications for Integrated Networks (JAIN) group, and the Parlay Group (now merged into the Open Mobile Alliance). In this book, we shall come across contributions from all of these groups.

2.2.1　IN Capability Sets

Many other advances were made in telecommunications at the time when ITU-T was standardizing IN. Telephony exchanges were becoming increasingly advanced and offered more and more programmable features. The Internet was expanding rapidly, connecting computers on a worldwide scale, and the World Wide Web was being born. Wireless operators were deploying the first digital wireless networks such as GSM and the digital advanced mobile phone service (D-AMPS). This made the standardization of IN a moving target.

To deal with the changing technological environment, ITU-T decided to standardize IN in a continuous process and publish stable specifications at regular intervals. ITU-T called these periodic releases of IN specifications capability sets (CS). As the name suggests, a capability set defines a set of service capabilities out of which services can be created. These capability sets are upward compatible; in other words, each new capability set adds new features while the existing ones remain valid.

The first capability set standardized in 1992 was CS-1. As IN was new at the time and there were few implementations around to validate the concept, it was necessary to polish the definition in 1995, resulting in the revised CS-1 (CS-1 R). Since then, the ITU-T has published CS-2 in 1997, CS-3 in 1999, and CS-4 in 2002. ITU-T has not published any IN standards beyond CS-4.

As explained in the previous section, ITU-T and U.S. standards differ both in terms of technical specifications and release planning. ANSI has also published the IN standards in releases, but these have a different timing than the ITU-T capability sets.

All IN capability sets are built on the same foundation, the IN conceptual model. The following section explains this model and how its structures the IN standards.

2.2.2　IN Conceptual Model

There are several ways to explain and understand IN. One of the ways to look at IN is as a collection of dedicated computers that perform special control functions for the telephone network. Another way of looking at IN is as a software architecture that runs applications to enhance the telephony service. In other words, one can define IN both from an infrastructure and from a software point of view.

The ITU-T standards combine these different views in an IN conceptual model (INCM) that reconciles the different physical and logical views of the

Table 2.1
ITU-T Recommendations That Define the INCM

Number	Recommendation
Q.1200	General Series IN Recommendations Structure
Q.1201	Principles of the IN Architecture
Q.1202	IN Service Plane Architecture
Q.1203	IN Global Functional Plane Architecture
Q.1204	IN Distributed Functional Plane Architecture
Q.1205	IN Physical Plane Architecture
Q.1208	General Aspects of the IN Application Protocol
Q.1290	Glossary of Terms Used in the Definition of IN

IN architecture. The INCM defines four different views of IN: the service plane (SP), the global functional plane (GFP), the distributed functional plane (DFP), and the physical plane (PP). At first sight these four planes seem to resemble protocol layers, but they are not. The INCM is just a convenient way of understanding IN from different angles. The INCM helps the network operator create new services top-down, first defining the high-level features that define the new service, then configuring the software building blocks that implement these features, and finally defining what nodes this software will be deployed on.

The INCM is defined in the ITU-T recommendations[3] Q.1200, Q.1201–Q.1205, Q.1208, and Q.1290, as shown in Table 2.1.

The INCM provides the structure for defining the capability sets. Each capability set is described in the same way: CS-x is described by recommendations Q.12x0, Q.12x1–Q.12x5, Q.12x8, and Q.12x9 with x = 1,2,3,… Table 2.2 shows the numbering scheme for IN CS-x recommendations.

In the following sections, we'll take a closer look at the different planes of the INCM, the foundation of all IN capability sets.

2.3 IN Service Plane

The service plane is the top plane of the INCM, describing services from the user's point of view. It describes what features a service is composed of but not

---

[3] ITU-T calls its standards "recommendations." Other standards organizations such as the 3GPP and ETSI speak of "specifications."

Table 2.2
Structure of ITU-T Recommendations Series Q.1200

Number	Recommendation
Q.12*x*0	Structure of IN CS-*x*
Q.12*x*1	Introduction to IN CS-*x*
Q.12*x*2	IN Service Plane Architecture for CS-*x* (not for CS-1)
Q.12*x*3	IN Global Functional Plane Architecture for CS-*x*
Q.12*x*4	IN Distributed Functional Plane Architecture for CS-*x*
Q.12*x*5	IN Physical Plane Architecture for CS-*x*
Q.12*x*8	IN Interface Recommendations for CS-*x*
Q.12*x*9	IN User Guide for CS-*x*

how these features are implemented in the network. A feature in the service plane can be seen as a reusable unit of functionality. A freephone service, for example, can be built from two IN features:

- *One number feature:* forwards calls to one single external number to the telephones of call center employees attending to the calls;
- *Reverse charging feature:* reverses call charges so that the owner of the freephone number pays instead of the caller.

Figure 2.7 illustrates the IN service plane concept.

The concept of services being composed of features is essential to IN. The key philosophy behind IN is that it standardizes reusable service features rather than services.

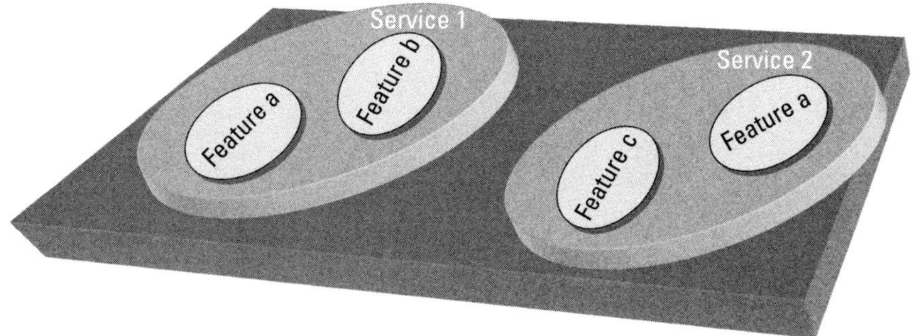

Figure 2.7 IN service plane.

Table 2.3
Features Defined in the CS-1 Service Plane

Feature Group	Features
Numbering	Abbreviated dialing, one number, personal number, private numbering plan
Routing	Call forwarding, follow-me diversion, time dependent routing, origin dependent routing, call distribution
Charging	Premium rate, reverse, split
Access	Authentication, authorization code, off-net access
Restriction	Call limiter, call gapping, closed user group, originating call screening, terminating call screening
Customization	Customer profile management, customer recorded announcement, customized ringing
User interaction	Originating user prompting, destination user prompting, attendant, consultation calling
Other features	Call waiting, automatic call-back, call hold with announcement, call logging, call queuing, call transfer, mass calling, meet-me conference

Table 2.3 shows the main features defined in CS-1. In this chapter, we will not discuss each feature in detail, since the idea is only to understand how the INCM works. Many of the features shown in Table 2.3 should look familiar though, as they are very similar to features found in private branch exchanges (PBXs).

Of course the available features determine to a large extent the kind of services one can create from them. The features defined for CS-1 are few and rather simple. They correspond closely to PBX services and to existing supplementary services that were programmed in telephone exchanges before IN was introduced, such as call forwarding. Capability sets CS-2, CS-3, and CS-4 offer additional and more advanced features that enable multiparty call handling, noncall-related user interaction (e.g., using the telephone screen), click to dial, and control of voice over IP calls.

2.4   IN Global Functional Plane

Whereas the service plane describes how services are composed from features from the user's viewpoint, the global functional plane (GFP) looks at services

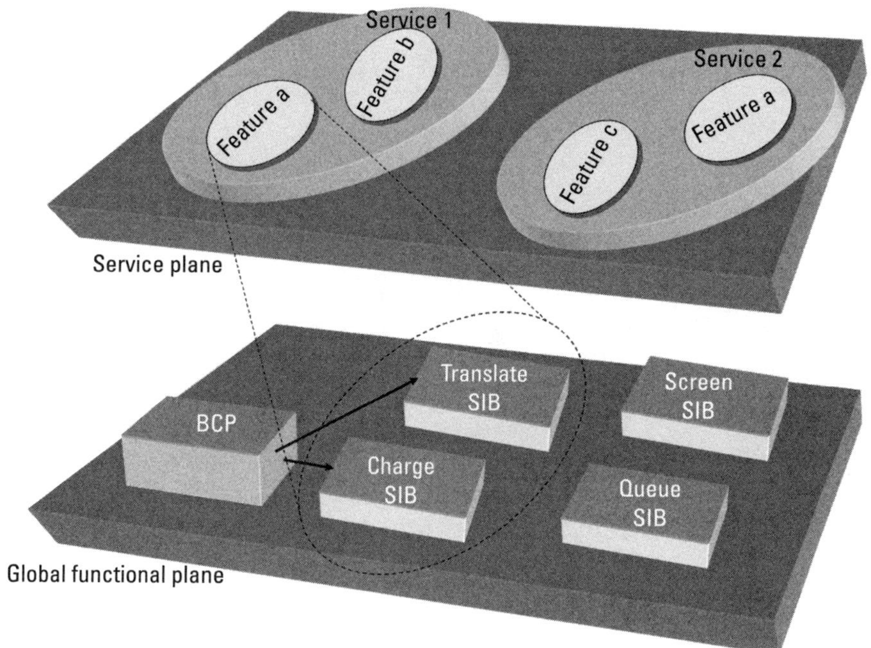

Figure 2.8　Mapping from service plane to global functional plane.

from the service provider's viewpoint. The global functional plane describes the software components that a service provider must deploy in order to assemble services. These components are called service independent building blocks (SIBs).

A feature defined in the service plane is implemented by one or more SIBs in the global functional plane. Figure 2.8 illustrates this relationship between the service plane and the global functional plane.

2.4.1　Service Independent Building Blocks

ITU-T did not define the GFP in terms of object-oriented programming, and SIBs are not objects. Keep in mind that object-oriented programming was only in its infancy and that Java didn't exist when IN was standardized in the first half of the 1990s. Also, telecommunications standards have to be universal specifications and usually avoid using specific programming languages such as C++ or Java.

SIBs are object-like software components that can be composed into service scripts. What mostly distinguishes SIBs from modern objects is that

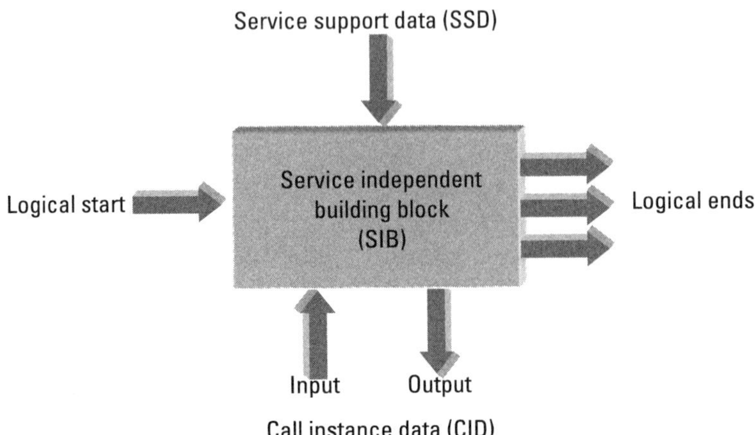

Figure 2.9   Service independent building block.

they explicitly show the control flow, allowing to be chained together as in a flow chart.

Each SIB has one logical starting point and several logical ends. The logical start and ends is where the SIB is connected into the control flow. The multiple logical ends allow the SIB to be used as a decision point in the service script.

The SIB acts on call instance data (CID). This is the call dependent data, like the originator's address, the dialed number and the destination routing address. Furthermore the SIB may need service support data (SSD), which is independent of the call. Examples of SSD are a call-screening list (black list), a call forwarding table, or a charging scheme. Figure 2.9 shows structure of a SIB with its parameters.

The global functional plane for IN CS-1describes a total of 14 SIBs for basic call handling. Later capability sets added only a few new SIBs for multiparty call handling, call unrelated services, and managing parallel SIB processes. Later capability sets also made the SIB concept recursive, allowing the SIBs to be composed into higher level SIBs that perform more complex functions. Table 2.4 summarizes the basic IN SIBs and their function.

As can be seen from this list, SIBs have simple functionality although the complexity does vary somewhat between them. For example, the user interaction SIB provides more complex functionality than the screen or algorithm SIB.

The IN standard describes for each SIB exactly what the logical start, the logical ends, the SSD, and CID are. Figure 2.10 shows a simplified diagram of the parameters for the screen SIB. This SIB can take the dialed number as CID input parameter and uses screening list (typically a black list or white list

Table 2.4
SIBs Defined in the CS-1 Global Functional Plane

SIB	Description
Algorithm	Applies an algorithm to the CID and/or SSD (the algorithm to be applied is specified as a parameter of the SIB).
Authenticate	Verifies a user's privilege to place/receive calls and to access particular service logic and/or data.
Basic call process	Basic SIB, which represents the setting up of a point-to-point call.
Basic call unrelated process	Describes user interactions that take place outside the context of a call (e.g., authentication, registration, location update).
Charge	Initiates special charging treatment (i.e., any charging that is different from the standard charging of a call by the SSP). The special charging treatment to be applied is a parameter of the SIB.
Compare	Performs a comparison of the input parameter against a specified reference value.
Distribution	Distributes the treatment of calls on the basis of specified parameters.
End	End a process.
Initiate service process	Create a new process, similar to doing a fork in Unix.
Join	Attaches or more call parties to a call.
Log call information	Writes detailed call information into a file. This information is intended for offline use, not within the context of the call.
Message handler	Send data from one process to another (when processing several SIB threads in parallel).
Queue	Queues calls for completion to a specific network address.
Screen	Compares the input parameter against a list of data.
Service data management	Enables the retrieval, storage, deletion, and modification of service data in the database.
Service filter	Accepts or rejects calls according to specified parameters. This SIB was originally called limit in CS-1 because it was mostly used for call limiting (to prevent network overloads).
Split	Detaches one or more call parties from a call and joins them in a new call.
Translate	Translates input parameter value into output parameter value, based on a specified algorithm or table.
User interaction	Allows information to be exchanged between the network and a call party (e.g., by playing an announcement or by receiving DTMF tones from a party).
Verify	Performs a syntactic check on the input parameter, according to a given syntax.

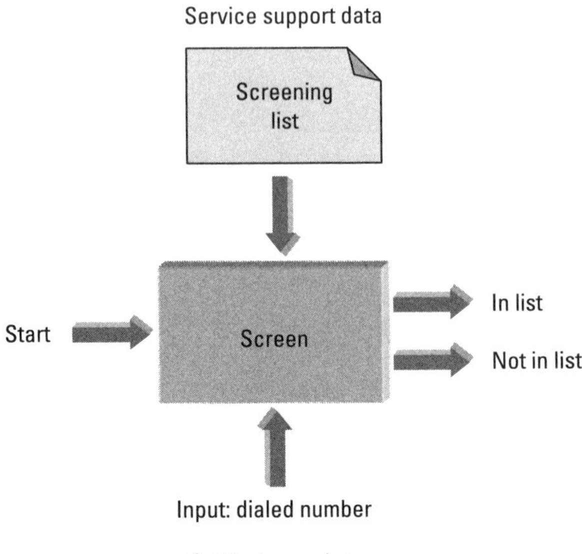

Figure 2.10   Parameters of the screen SIB.

of telephone numbers) as SSD against which to check the number. Depending on the outcome of this check, the screen SIB returns either at the in-list or the not-in-list logical end.

The screen SIB is typically used to implement services like originating call screening (OCS) and terminating call screening (TCS). OCS screens the dialed number at the originating end of the call (as in the screen SIB of Figure 2.10). This service is mostly intended for limiting international calls or calls to premium rate numbers from certain telephones.

TCS screens the identity of the calling line at the receiving end of the call. This service is typically used to block unwanted incoming calls (e.g., malicious calls or calls from telemarketers).

Figure 2.10 shows only a simplified diagram, and in reality the parameters are more complex. The logical ends usually also include an "error" clause for the cases where there is a syntactical error in one of the inputs. For the sake of simplicity, Figure 2.10 does not show such detail. The important thing for now is to understand the mechanism by which the SIBs are defined.

2.4.2 Basic Call Process

Among the SIBs, there is one special building block that represents the call itself. This special SB is called the basic call process (BCP). The BCP describes the phases of setting up a telephone call from one user to another. At each of these phases, it is possible to interrupt the call setup process and to start executing a service program.

The points where the call processing can be interrupted are called points of initiation (POIs). When a service is started at a POI, we say that the service is triggered; for this to happen, it is necessary to arm a trigger at the POI beforehand so that the BCP knows it must suspend processing.

At the time of triggering, the BCP passes the CID to the first SIB in the chain to be executed. The SIBs that make up the service can then change the CID as required by the service. In the case of call forwarding, for example, they can change the destination address.

After finishing the processing of the service, the last SIB hands back control to the BCP. The service can order a jump to any point in the BCP call processing; it does not have to be at the same point that the service was invoked.

The point where control is handed back to the BCP is called the point of return (POR), and at this point the BCP can be instructed to handle the call in a specific way. Table 2.5 show the POIs (on the left) and the parameters that can be passed to the BCP with the POR (on the right), as they are defined in CS-1. The later capability sets specify a number of additional POIs and PORs for intervention during the active state of the call, such as a hook flash or pushing the special R button on the phone.

To illustrate how a service is described at the global functional plane, let's consider the example of a call made with a calling card. Figure 2.11 shows how this service could be implemented with three SIBs. The example is, of course, oversimplified and only serves to help understand how services are

Table 2.5
POIs and POR Parameters of the CS-1 BCP

Points of Initiation (POIs)	Points of Return (PORs)
Call originated	Handle as transit
Address collected	Continue with existing data
Address analyzed	Proceed with new data
Call arrival	Provide call party handling
Busy	Initiate call
No answer	Clear call
Call acceptance	
Active state	
End of call	

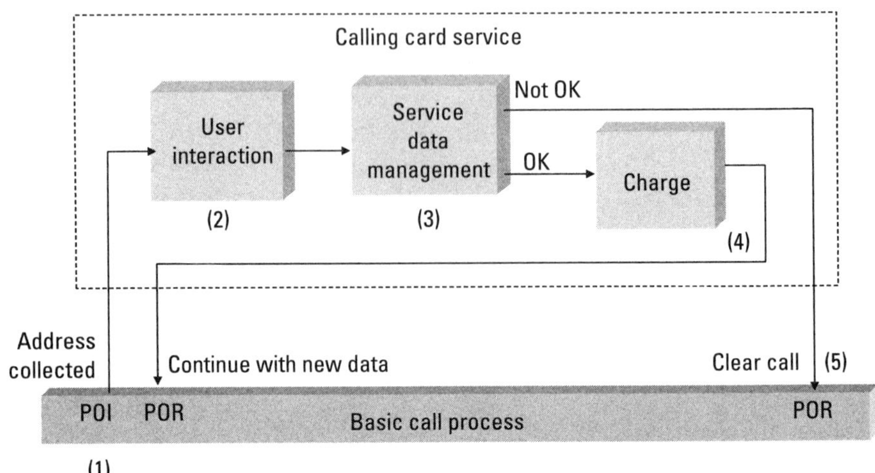

Figure 2.11   Calling card call service defined at GFP level (simplified).

defined with SIBs. Figure 2.11 shows only the logical flow of SIBs. All SIB parameters (CID and SSD) have been omitted for simplicity.

According to Figure 2.11 the service could be implemented as follows. Imagine that the user dials a special number corresponding to the calling card call service:

1. At the address collected POI the BCP immediately recognizes that the dialed number is special and triggers the calling card call service.
2. The first SIB to be executed is the user interaction SIB. This SIB plays a message to the user: "Please enter your card number, PIN code, and number you want to dial." It collects the DTMF tones from the user and converts them to numbers.
3. The next SIB in the sequence, service data management, uses these numbers to look up the calling card for this user and check the PIN code.
4. If the card exists and the PIN code is correct (the "OK" branch), then the charge SIB starts charging the call to this calling card. After this, control is passed back to the BCP at the same point where call control was suspended. The POR carries the instruction to continue with new data, causing the call to proceed with the destination number input by the user.
5. If the card number or the PIN code is incorrect (the "not OK" branch), then the service instructs the BCP to terminate the call immediately by jumping to the clear call POR.

This example is, of course, strongly simplified. Real-life services can take up to hundreds of SIBs, although most of the service logic is usually for error handling. Services like the calling card call have to take into account any possible error: wrong card number, incorrect PIN, invalid destination number, time-out of user response, and so on. Service providers say that as much as 80% of the service logic can be spent on error handling.

2.5 IN Distributed Functional Plane

The GFP helps a service designer to identify the building blocks out of which to build services, but its model of the network is too abstract for implementation. In reality we cannot manage the network as if it were one machine with a single BCP for each call. In practice, the network has a complex layered structure and consists of many specialized elements. Things get even more complicated if we think about calls that start in one operator's network and end in another's. The GFP does not provide enough technical detail to actually implement the services in the network.

The purpose of the distributed functional plane (DFP) is to provide a more realistic view of the network that reflects the distribution of functions. In reality the setup of a telephony call is not a centralized function, as the GFP assumes. It is the result of interactions between network nodes that exchange messages to decide how to switch the call from source to destination. To reflect this reality, the DFP distinguishes different functions in the network.

2.5.1 Functional Entities

The DFP identifies the different functional entities (FEs) that make up the telephony network. FEs are distributed network functions that interact to run the BCP and the other SIBs defined at the GFP. A single SIB can require more than one FE for its execution.

The DFP defines what information has to be exchanged between the different FEs to implement each SIB. These information flows between FEs are specified in the specification and development language (SDL)[4] and with

---

[4] SDL is a formal specification language defined by ITU-T in the recommendations Z.100 series. Formal specification language describes models rather than executable code and are used mostly in system design and validation prior to implementation. SDL has a graphical format that allows designers to describe systems in terms of communicating processes. Processes are defined by their states and transitions between them, and have formally defined semantics in terms of state machines. SDL also has a textual format to allow tools to import and export specifications. SDL

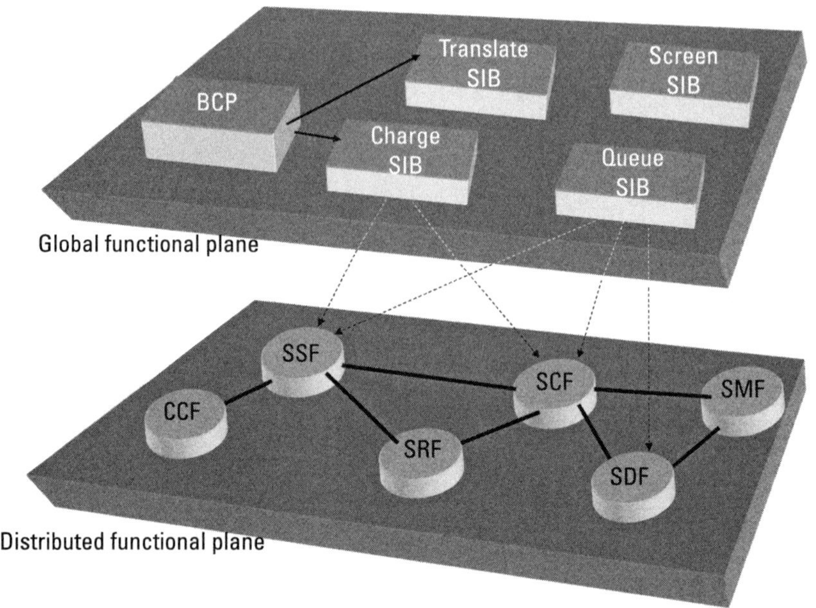

Figure 2.12 Distributed functional plane.

message sequence charts (MSCs).[5] Both are ITU-standardized languages for the formal description of distributed systems.

Figure 2.12 illustrates how the implementation of SIBs is distributed over FEs.

The most important FEs defined in CS-1 are as follows:

- *Call control function* (CCF): keeps the state of a call (e.g., "alerting" or "terminating line busy").
- *Service switching function* (SSF): takes care of controlling the bearer connections in a call, typically voice circuits.

---

is aligned with the unified modeling language (UML), allowing designers to specify systems with combined UML and SDL diagrams.

[5] Message sequence charts provide a way of describing the message flow in a system. They consist of vertical lines, with horizontal arrows connecting pairs of vertical lines. Each vertical line represents a component or object and represents a time axis with time flowing from top to bottom. An arrow between one vertical line and another represents a message flowing from one component to another, at the given point in time. This allows designers to describe how components exchange messages in temporal order.

- *Service control function* (SCF): contains the programs that control services. Such programs are called service logic in IN.
- *Service data function* (SDF): represents the databases that keep the service support data (e.g., number translation tables or prepaid account credit data).
- *Service management function* (SMF): contains all management functions, such as installing new services, adding or deleting service subscribers, and updating service data.
- *Special resource function* (SRF): represents special functions in the network that could be needed to realize services, like playing announcements, playing tones, generating special ringing patterns, or bridging and mixing of conference calls.

The difference between the CCF and SSF deserves some explanation. It is natural to wonder why there are two separate functions to represent the setting up of telephone calls. From the user perspective, the call and the connection are the same thing: wouldn't it be sufficient to have just the SSF to represent the functionality of the exchange?

Remember that SS7 already separates signaling from connections. In other words, the signaling involved in setting up a call is carried on a different network than the voice connections themselves. Taking this a step further, we could imagine a call as being not just a connection from A to B, but a general concept that associates parties (users placing or receiving the call) with connections and network resources during a delimited period of time.

There are important advantages to separating the concept of a call and a connection. It allows for the concept of a call to extend naturally to conferences, multimedia communications, and even Internet connections. And it allows service logic programs (like IN) to control the adding and dropping of connections within the context of a single call. It is especially for this reason that IN separates the CCF from the SSF.

CS-1 requires that a service be controlled by a single SCP, as was the practice in Bellcore's original AIN and early IN products. CS-2 and subsequent capability sets lifted this restriction and allowed services to be controlled by two or more communicating SCPs, distributing the service execution over several nodes. This made it possible to control services across service provider domains, at least in theory. In practice, the level of coordination and agreement needed between two (often competing) service providers to have their SCPs support the same distributed service logic often proved to be too much of a barrier to providing cross-provider IN services.

Later capability sets also defined additional FEs. Two of these, defined in CS-2, handle user interactions that take place outside a call:

- *Call unrelated service function* (CUSF): processes events from call-unrelated user interactions and can trigger service processing in the SCF. The CUSF is like the SSF, but for call-unrelated events.
- *Service control user agent function* (SCUAF): represents the user interfaces that communicate with the CUSF. Although the SCUAF does not have a real counterpart in CS-1, one could argue that its function is similar to that of the SRF but for call-unrelated interactions.

Two other FEs added CS-2 allow IN FEs to interwork with non-IN networks, and in particular voice over IP equipment and websites:

- *Service management access function* (SMAF): provides a user interface for external access to the SMF (e.g., through a website).
- *Intelligent access function* (IAF): an interworking function that allows an SCF to control non-IN equipment such as voice over IP gateways. We'll talk about voice over IP and soft switching in later sections.

Some FEs introduced in CS-3 for interworking with mobile networks were never really applied in practice. These include the call-related radio access control function (CRACF), call unrelated radio access control function (CURACF), and radio control function (RCF). Rather than using these FEs, mobile networks tend to use an adapted version of IN called CAMEL, which we'll learn more about later in this chapter.

2.5.2 Half Call

The DFP provides a more detailed representation of the basic call process. The most important refinement is that the DFP takes into account the distributed nature of a call. At the DFP level, the call is no longer seen as a single, abstract process, but distinction is made between call processing for the originating (calling) party and for the terminating (called) party.

The DFP introduces the concept of a half call: the BCP is now split in two parts, called the originating basic call state model (O-BCSM) and the terminating basic call state model (T-BCSM). The O-BCSM describes the procedure for initiating a call, while the T-BCSM describes the process of receiving one. Both the O-BCSM and T-BCSM can trigger services in the SCF, as Figure 2.13 illustrates.

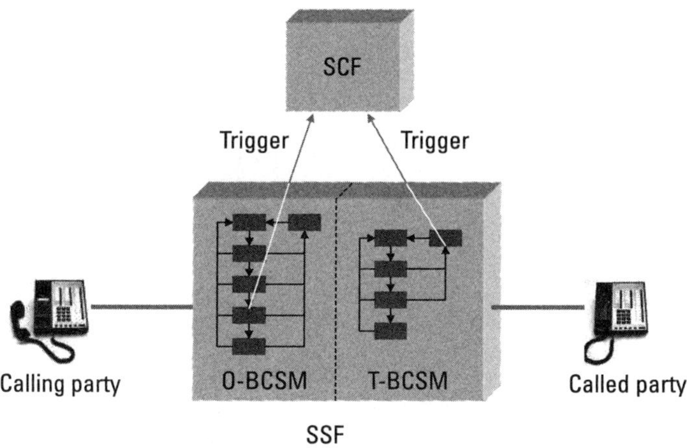

Figure 2.13 The half call concept as defined in the DFP.

It is obvious why the DFP distinguishes between O-BCSM and T-BCSM: there exist IN services that deal with outgoing calls (e.g., abbreviated dialing) and services that deal with incoming calls (e.g., blocking of malicious calls). Apart from this, the caller and called party are not necessarily connected to the same exchange or even to the same network, so one can't assume that there is one global process that has control over the whole call.

2.5.3 Detection Points

The O-BCSM and T-BCSM can both be seen as more detailed versions of (half of) the BCP at the GFP. Both the O-BCSM and T-BCSM are represented as state-transition diagrams that describe the call processing states and transitions between them. The states are called points in call (PICs), and the transitions are caused by events. Examples of events are taking the phone off the hook, dialing a number, putting the phone on the hook, or flashing (swiftly pushing) the hook.

An event can have a detection point (DP) associated with it. DPs can be seen as implementations of the POIs defined at the GFP: they are points where the SSF can suspend call processing and hand over control to the SCF. Figure 2.14 shows the PICs and DPs of the O-BSCM and T-BCSM as defined in CS-1.

Each large dark box in Figure 2.14 represents a PIC. The lines and arrows represent events, and the small darker boxes represent DPs. The names of the PICs and DPs are mostly self-explanatory. In fact the IN specifications prefix all names in the O_BCSM with O_ and all names in the T-BCSM with T.

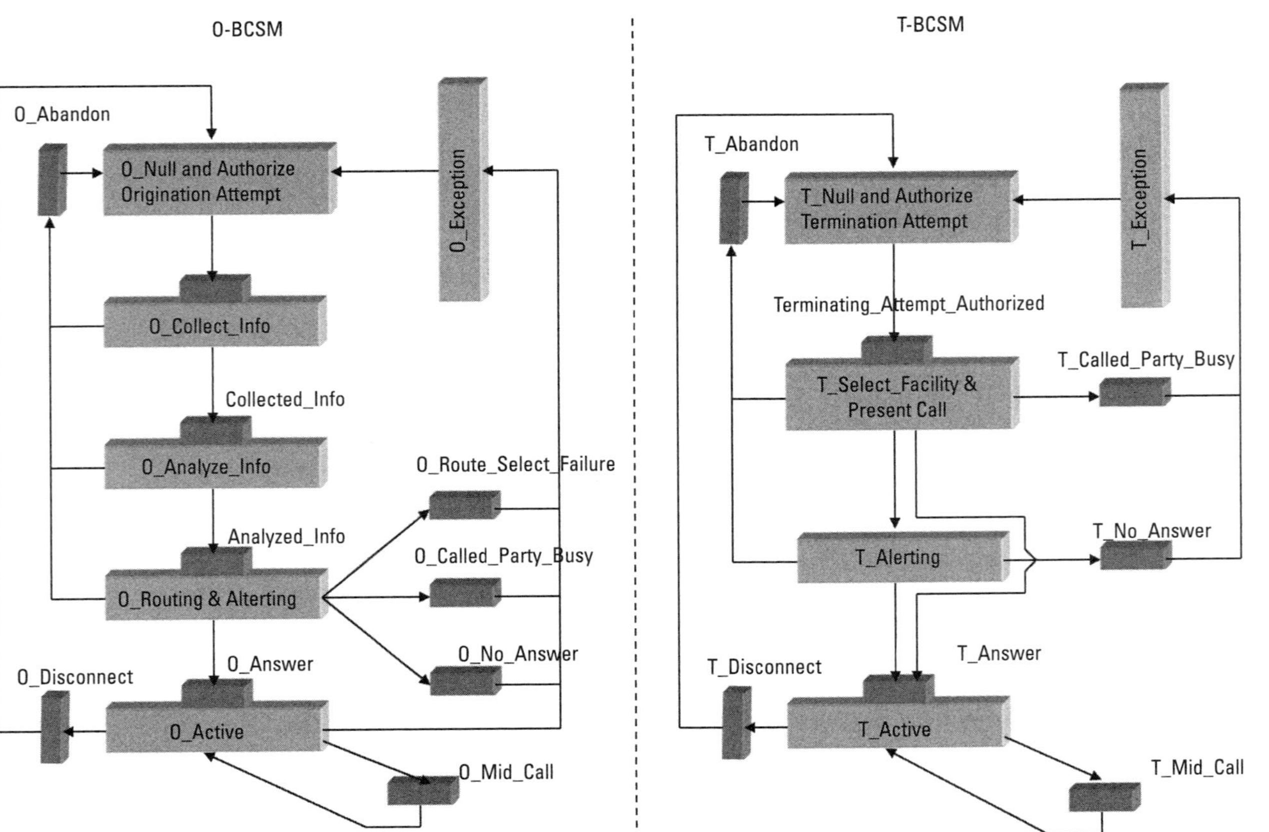

Figure 2.14 Basic call state models for originating and terminating side.

CS-2 defines a more detailed O_BCSM and T_BCSM by breaking down some of the PICs into call states of a finer grain. Table 2.6 shows how CS-2 expands CS-1 PICs; the CS-1 PICs not mentioned in the table remain unchanged in CS-2. CS-2 also defines two new PICs, which represent a suspended call that can be resumed later, O_Suspend and T_Suspend. Of course, with the new PICs in CS-2 also come a few new DPs.

CS-2 also defines an additional basic call unrelated state model (BCUSM) that represents user interactions taking place outside the context of a call. The BCUSM has a simple structure and is drawn in Figure 2.15. The states and transitions of the BCUSM represent the activation and release of a signaling channel for user interaction, and the reception and sending of user data in the form of components. All this takes place outside the context of a call.

A DP can take on two forms, depending on how it is activated:

- *Trigger detection points* (TDP): set statically at the time of deployment of a service, or when a user subscribes to it;
- *Event detection points* (EDP): set dynamically by service logic during the course of a call.

Table 2.6
Comparison of CS-1 and CS-2 PICs

CS-1 PIC	CS-2 PIC
O-BCSM	
O_Null&Authorize_Origination_Attempt	O_Null
	O_Authorize_Origination_Attempt
O_Routing&Alerting	O_Route Select
	O_Authorize_Call_Setup
	O_Send Call
T-BCSM	
T_Null&Authorize_Termination_Attempt	T_Null
	T_Authorize_Termination_Attempt
T_Select Facility&Present Call	T_Select Facility
	T_Present Call

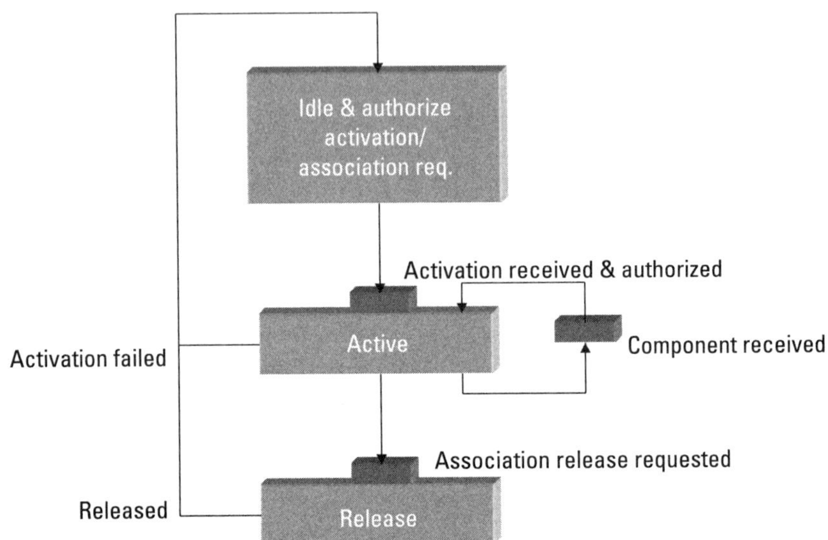

Figure 2.15   Basic call unrelated state model.

For example, one would use a TDP for a service like freephone, where the DPs are only activated when the service is deployed, but you need an EDP to realize a service like automatic callback on busy where the DP has to be activated during the call.

When the O-BCSM or T-BCSM encounters a detection point there are two kinds of action possible:

- *Notification:* the BCSM only sends information to the SCF and continues without waiting for a response.
- *Request:* the BCSM suspends call processing, hands control to the SCF, and awaits a response before continuing, possibly with modified call parameters.

The call forwarding service, for example, can be started only as a request because call processing must wait for the number to be translated. On the other hand, a televoting service may only need information to be sent to the SCF without suspending the call.

Depending on whether a detection point is a TDP or an EDP, a notification or a request, there are four possibilities, as shown in Table 2.7.

Table 2.7
Detection Point Types

	Trigger Detection Point	**Event Detection Point**
Notification	Trigger detection point-notification (TDP-N)	Event detection point-notification (EDP-N)
Request	Trigger detection point-request (TDP-R)	Event detection point-request (EDP-R)

2.5.4 Trigger Processing

The SSF must have a mechanism for handling DPs when they are encountered during the processing of a call. In practice this is usually a table that lists which DPs are armed and whether they are TDP-N, TDP-R, EDP-N, or EDP-R. The IN standards also define a function in the SSF that detects and resolves conflicts between different services triggered on the same call, the feature interaction manager (FIM).

Figure 2.16 shows an example of the DP processing mechanism in the SSF. It doesn't really matter in this example whether we're looking at an O-BCSM or T-BCSM, because DP processing is the same for both cases.

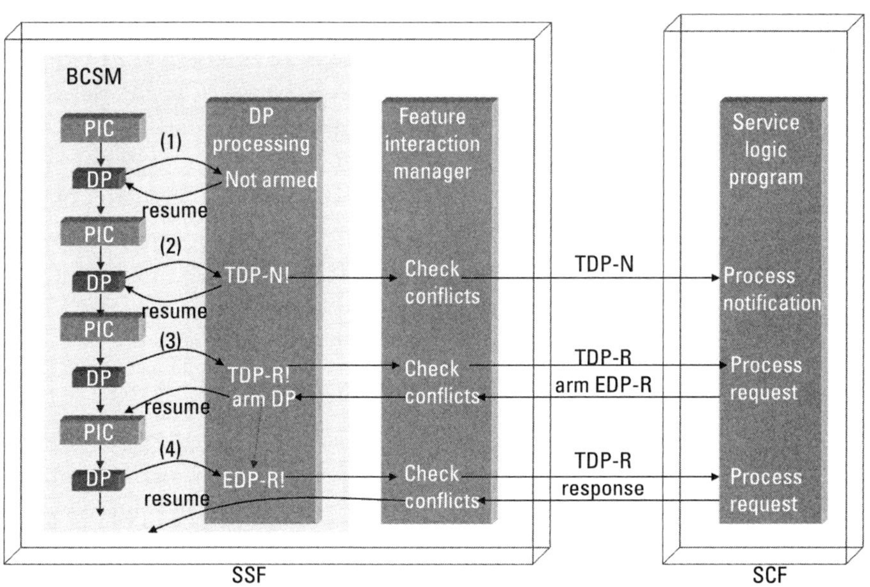

Figure 2.16 DP processing and feature interaction manager.

Figure 2.16 shows the following sequence of events and triggers:

1. The first DP encountered by the BCSM is not armed, and the BCSM is instructed to continue.
2. The next DP is armed as TDP-N, meaning that only a notification is sent to the SCF and the BCSM immediately resumes without waiting for a response. Before relaying the TDP-N to the SCF, the FIM checks whether there are any interactions with other services activated on the same call.
3. The third DP encountered is armed as a TDP-R. This time it is a request, so call processing is suspended and the request is sent via the FIM to the service logic. In this example, the service logic responds with a command to arm an EDP-R further on in the BCSM. This command is passed back to the DP processing module, which takes care of arming the EDP-R. Then it instructs BCSM to resume.
4. The last DP encountered in this example is the one previously armed by the service logic. This is an EDP-R, which means that call processing is also suspended until the SCF returns a response (e.g., a translated number).

In this example, the BCSM resumes at the next PIC. It is important to note that the service logic may instruct an event to be armed at any DP in the BCSM and that call processing can resume at any point in the BCSM. In other words, you don't always have to resume to the next PIC. For example, in a service where you check a PIN code, it makes sense to jump directly to the end of the BCSM and release the call if the PIN is entered wrongly.

From this section, it is obvious that the distributed functional plane is significantly more complex than the global functional plane, but also more realistic in its modeling of call processing and the ways in which control can be passed to the SCF.

2.5.5 Call Party Handling

CS-1 only allows control of calls from one party to another and considers the call and the connection between the two parties to be the same thing. This means that CS-1 has trouble controlling calls that involve multiple parties that can join and leave during the call, as is common in conference calls. Even a relatively simple service such as call waiting is difficult to realize with CS-1 because it requires a call to control three connections rather than just two.

CS-2 makes it possible to add, remove, and reroute individual parties and their connections within a call. In other words, CS-2 distinguishes the call from the connections it manages. This important CS-2 feature is referred to as call party handling (CPH).

The addition of CPH makes CS-2 and later capability sets much more complex than CS-1. The GFP does not really reflect this complexity: there are only two new SIBs: split and join. But at the DFP the complexity becomes visible: CHP requires the call and its connections to be processed separately and requires additional network equipment such as conference bridges to connect multiple connections together.

One of the most important requirements is that the service logic in the SCF must always have a consistent view of the configuration of the connections in a call. When the SCF loses track of who is connected to the call, the service gets into an inconsistent state and the effect of a join or split operation becomes unpredictable. For this reason CS-2 defines a formal model for describing the state of connections in a call. This is called the connection view state (CVS) model.

The main elements in a CVS are as follows:

- *Call segment* (CS) describes a half call as explained in Section 2.5.2, but in CS-2 a half-call can also apply to multiparty calls. A CS is described in terms of legs and connection points.
- *Connection point* (CP) is a joint between two or more legs, or between just one leg and the network. The notion of a CP is generic: it can be a switched connection but it can also be a conference bridge or a media distribution point.
- *Leg* represents a part of the communication path between parties in a call. A leg can be *controlling* or *passive*, depending on whether service logic was invoked for this leg or whether it just receives instructions passively.
- *Call segment association* (CSA) is the group of all call segments that are involved in a given service and that are controlled by the SCF. In principle a call segment can be made out of any combination of legs and connection points, but not all combinations are meaningful. CS-2 identifies 14 specific configurations that are being found in the most common services.

The CS-2 standard uses a graphical notation to describe CVS. The diagrams for call segments show a connection point and the legs attached to it.

The CVS shows whether each leg is controlling (indicated by the letter c) or passive (indicated by the letter p). The CVS can also describe whether the leg has actually been set up (called *joined*) or whether it is still in the phase of being set up (called *pending*).

To show how the CVS represents the state of the connections in a call, let's consider the very simple example of the follow-on call feature. This feature is used, for example, in credit card calls and allows a user to make several calls without putting the telephone on-hook and entering the credit card details every time. Figure 2.17 shows a simplified version of the information flows between the SCF and SSF for the follow-on call feature, and the resulting CVS.

The information flow for the follow-on call leads to the following connection view states:

1. When the user places the first call, the O-BCSM in the SSF detects a trigger and sends the initialDP message to the SCF.
2. The SCF instructs the SSF to connect the originating and destination parties, and requests the SSF no report any midcall events.
3. When the user ends the first call and wishes to make a follow-on call, he or she does not put the telephone on-hook, but simply flashes the hook. This causes a midcall event to be reported to the SCF.
4. As a result, the follow-on call service logic does not terminate the call, but disconnects only the leg to the destination party.
5. Next, the SCF instructs the SSF to establish a connection to the SRF.
6. The SCF instructs the SRF to play an announcement message and prompt the user to dial the number for the follow-on call.
7. When the SCF has received the digits for the next call, it orders the disconnection from the SRF.
8. Finally the SCF sets up the connection to the new call destination specified by the user.

This example was taken in simplified form from recommendation Q.1229 (see Table 2.2 for the structure of ITU-T recommendations). The actual INAP messages between the SSF and SCF are more complex than shown in Figure 2.17, but what is important is to understand the principle. The CVS represent how the SCF sees the connections in the SSF. When a command from the SCF to the SSF changes the connections in a call, the CVS change accordingly.

The CVS can become very complex, and in practice it is very difficult to read and interpret them in the standards. This is because IN CS-2 tries to control multiparty communications with classical switching technology optimized for two-party telephone calls. This is like trying to map a complex

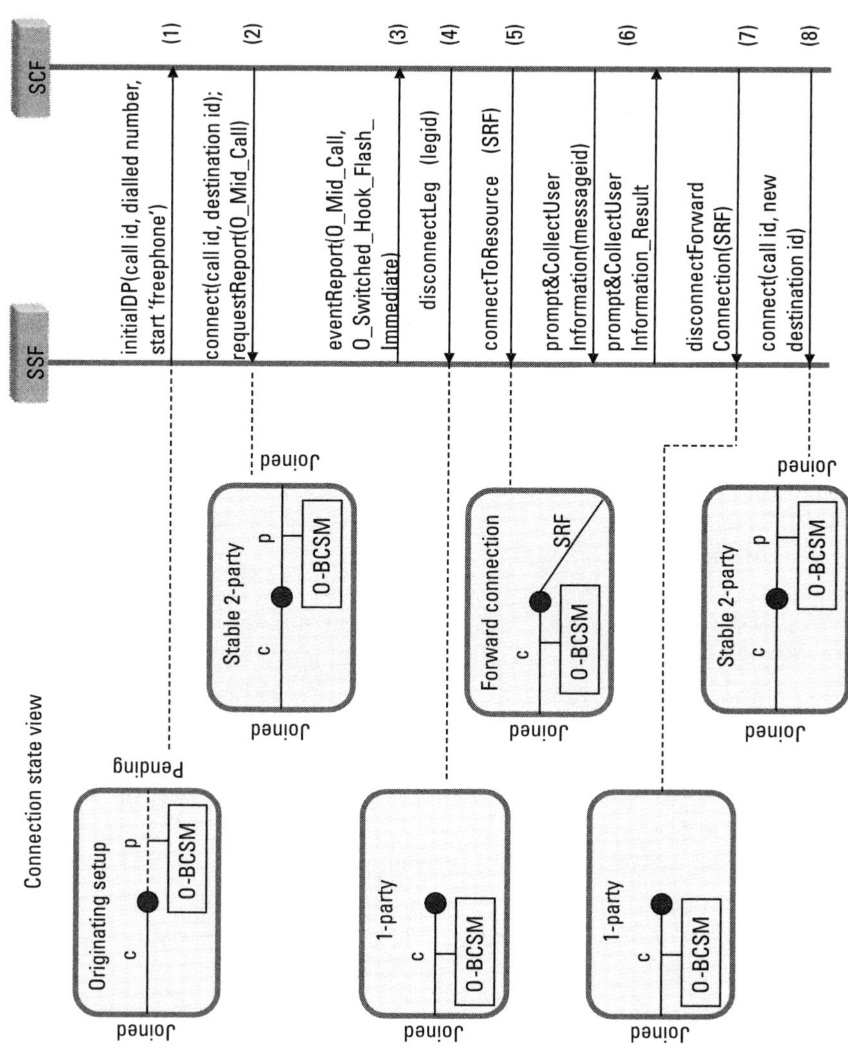

Figure 2.17 Connection state views for follow-on call.

three-dimensional model to a two-dimensional picture without losing any detail.

2.6 IN Physical Plane

The DFP takes into account the distributed nature of the network, but only in a superficial way. The GFP recognizes the difference between the calling and called party, and between call control and connection switching, but does not go as far as allocating functions to physical locations or machines. The physical plane (PP) provides the view of the actual network nodes and messages flowing between them.

2.6.1 Physical Entities

The physical plane allows for different mappings of IN functions onto physical nodes. A straightforward mapping simply assigns each function to a corresponding network node:

- *Service switching point* (SSP): represents the exchange and typically hosts both the CCF and SSF;
- *Service control point* (SCP): hosts the SCF;
- *Service data point* (SDP): hosts the database or databases that make up the SDF;
- *Service management point* (SMP): hosts the SMF;
- *Intelligent peripheral* (IP): implements the SRF.

Figure 2.18 illustrates this straightforward mapping of functional entities onto physical ones.

The mapping of Figure 2.18 would require at least four or five machines to implement an IN system. In smaller networks or in cases where there are only few services or few service subscribers, the number of machines can be reduced by collocating certain DFP functions onto the same physical node. For performance reasons, the SCP very often combines the SCF and SDF, as shown in Figure 2.19.

The configuration requiring least hardware is one where the SCF, SDF, and IP are all located together on one machine. Such a configuration is often referred to as a service node. Many vendors actually combine these functions with the exchange itself, and in this case the service node even includes the SSF. Figure 2.20 illustrates the service node concept.

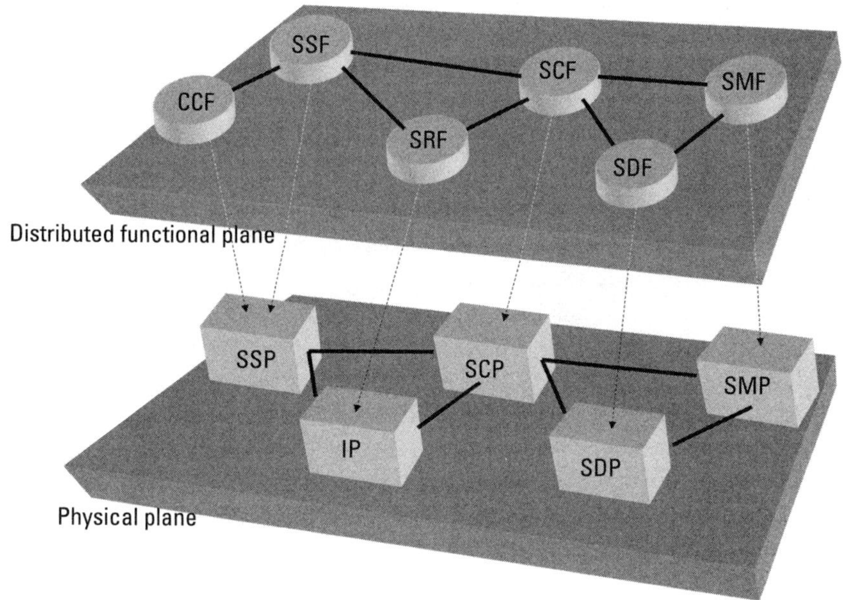

Figure 2.18 IN physical plane.

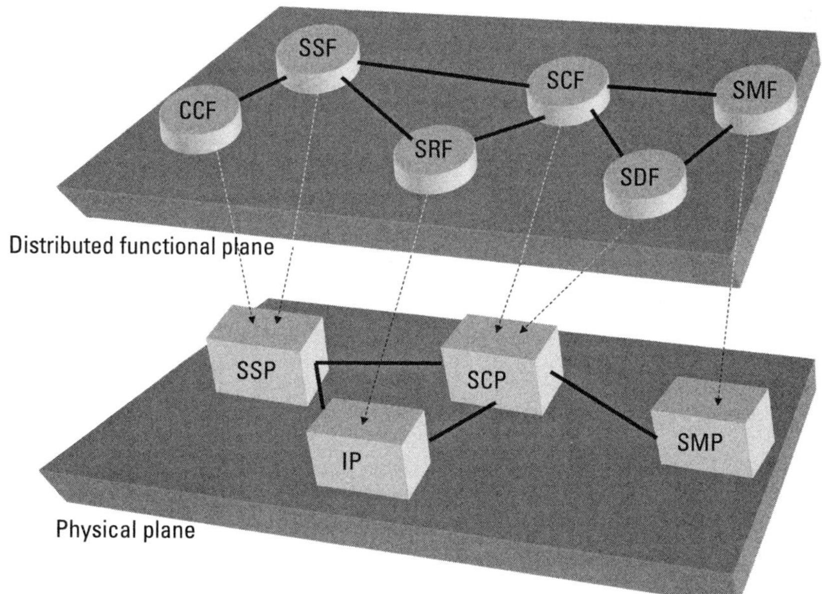

Figure 2.19 Collocation of SCF and SDF at the physical plane.

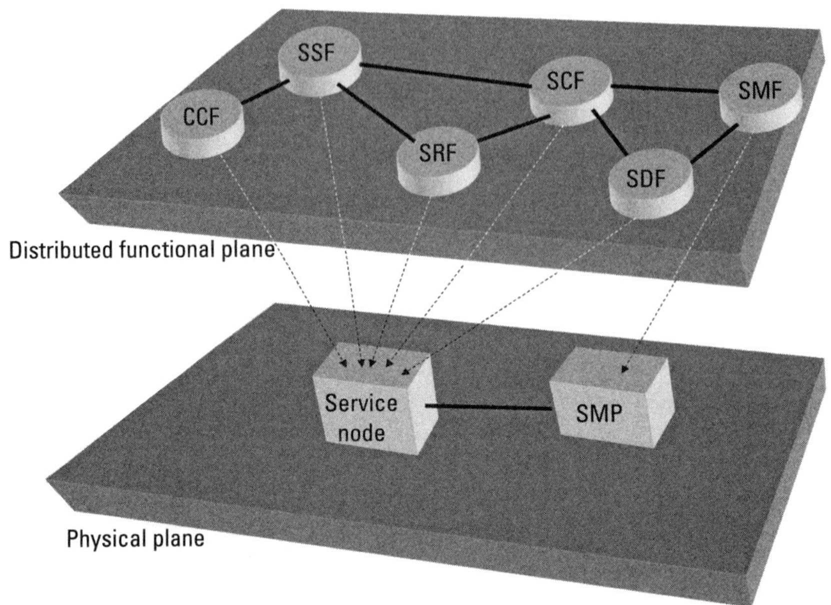

Figure 2.20   Collocation of all IN functions in a service node.

In practice, one can find all of these configurations on the market. Service nodes tend to be used in smaller networks or in networks with few value-added services. Nevertheless, service nodes represent an important market, and some vendors concentrate completely on this business segment.

Many manufacturers support a range of configurations and adapt their offer to the operator's requirements and budget. The INCM allows IN to be built in a modular way by creating a software module for each FE. This allows the manufacturer to distribute FEs over physical machines according to the customer's needs.

2.6.2 IN Application Protocol

The INCM describes the FEs that make up the DFP and the physical entities in the PP to host them, but leaves one important question unanswered: how do these functions and nodes interact? What messages do they exchange?

The IN application protocol (INAP) specifies the information exchanged between functional entities, most importantly between the SSF and SCF, and between the SCF and SRF. The communication between the SCF and SDF is a special case, since it is not defined by INAP but uses the directory access protocol (DAP) also specified by the ITU-T in its X.500 recommendations.

INAP defines only IN-specific messages and conveniently uses the SS7 network for transporting them between network nodes. Remember that SS7 was designed in such a way that it can transport messages of other protocols within its TCAP messages. This is why INAP is called an application protocol: from the SS7 point of view, IN is a telephony application.

Strictly speaking, ITU-T does not define INAP as part of the physical plane, but in separate documents (Q.1208, Q.1218, and Q.1228). These documents specify INAP using the abstract syntax notation 1 (ASN.1). ASN.1 is a complex language and difficult to read, but it is formal and unambiguous, and mostly used for machine processing rather than human enjoyment. The standard is very precise about the data types exchanged (e.g., a string of digits is not the same thing as a number!).

INAP is not just a collection of stand-alone messages; the sending of messages must follow certain order, called a *dialogue*. These dialogues implement the information flows defined in the distributed functional plane. Figure 2.21 shows a very simple example of such a dialogue for the freephone service.

1. When the BCSM recognizes the number dialed by the user as a freephone number, this triggers a TDP-R.
2. As a result, the SSF sends an initialDP message to the SCF with the calling line identifier, dialed number, and service identifier as parameters. The meaning of the initialDP message is to start the service processing in the SCF. As explained before, a service is defined as a sequence of SIBs in the GFP. In the DFP these SIBs map to programs in the FEs that communicate by sending messages to each other.
3. The service program in the SCF first queries the database in the SDF to find out where to route this call. It does this by sending a requestInfo message to the SDF.
4. The SDF looks up the dialed freephone number in the database, finds the corresponding destination network address, and returns it to the SCF in a requestInfoResponse message.
5. The SCF then orders the SSF to connect the calling party to the destination address, using the connect message. The SCF also requests the SSF to be notified when the connection is actually established and stopped. For this it sends a requestReport message to the SSF, indicating the DPs where it wants to be notified (in this case, O_Answer when the call is accepted and O_Disconnect when the call is disconnected).

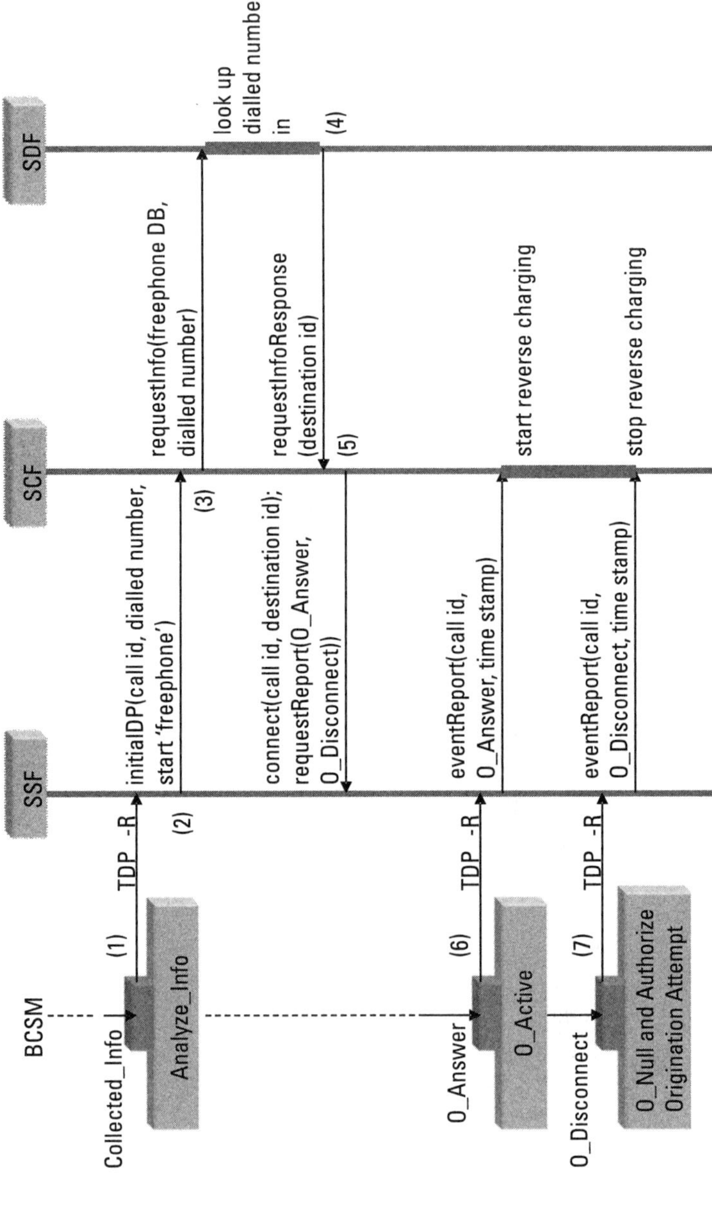

Figure 2.21　Simplified INAP dialogue for freephone service.

6. When destination party answers the call, the SSF sends an eventReport message with a time stamp. As a result, the SCF starts reverse charging the call to the destination address (remember, this is a freephone service so the caller doesn't pay!).
7. Finally, when one of the parties releases the call, the SSF sends another event Report message to the SCP, causing the SCP to stop charging and terminate the freephone service processing.

This example is, of course, strongly simplified so as to keep it comprehensible. In reality, the dialogues are much longer, and the messages have many more parameters. In fact the INAP specifications are hundreds of pages long and grow with each capability set. But what is important here is to understand the basic mechanism: dialogues of INAP messages implement the information flows between FEs, which in turn implement the SIBs, the building blocks out of which services are made.

Connecting different IN systems critically depends on their ability to exchange INAP messages. In theory, one can connect the use of an SCP from one vendor to the SSP from another as long as both strictly adhere to the INAP standard. In reality, many vendors have taken the liberty of using their own INAP subsets and dialects, which hampers interworking with other vendors' equipment. In some cases, this has been the purpose, so as to avoid network operators from purchasing equipment from the competitor. In other cases, it was due to the sheer complexity of INAP, prompting vendors to use simplified, more manageable message sets.

ETSI recognized this problem and standardized core INAP, a simplified version of INAP stripped of unnecessary complexity and a kind of common denominator of the INAPs in the market. Today, many manufacturers have core INAP-compatible IN systems. ETSI also sharpened the SDL specifications of the DFP to the point where they are formal enough to allow for the automatic generation of test scenarios. ETSI used the SDL models to specify protocol implementation conformance statements (PICS) that can be used as formal references for IN implementations. With this formal approach, ETSI also revealed and resolved a number of ambiguities in the original ITU-T standards. Thus, although ITU-T initiated the global standardization of IN, it is fair to say that ETSI made decisive contributions to IN in the second half of the 1990s.

Having seen the SP, GFP, DFP, PP, and INAP we have completed our tour of IN as standardized by ITU-T and ETSI. Next we will see how IN had to be adapted to fit mobile networks as they were emerging in the 1990s.

2.7   IN in Mobile Networks

A large part of IN standardization efforts took place in the 1990s, at the same time that mobile networks and the Internet were expanding. When IT published the first IN standards, mobile operators started rolling out second generation (2G) digital mobile networks on a wide scale. Soon operators and manufacturers were thinking of ways to apply IN to digital mobile networks such as GSM, D-AMPS, and IS-95. This would allow them to provide the same kind of services in mobile networks that they were providing in fixed telephony networks. Probably the most compelling and urgent case for operators was to enable prepaid calling, because prepaid subscribers turned out to be a big market in mobile networks.

At first sight it appears as though mobile networks can use IN in a straightforward way. After all, they are switched networks just like the telephone network. Unfortunately things are not quite that simple because mobile exchanges have to do more than switching alone: they also manage the mobility of terminals and subscribers. The next section will shed some more light on how mobile networks have to manage mobility and how that affects call and connection processing in mobile exchanges.

2.7.1   The GSM Network

GSM is the most widely deployed 2G digital mobile network, which relies on circuit switched connections. Although there are other mobile network standards such as D-AMPS and IS-95, they differ mostly in the radio technology they use. Their mobility management procedures are very similar, so we'll use GSM as the leading example here.

GSM is a cellular network. It provides radio coverage with many radio cells of a limited range rather than through a few big antennas. The use of cellular coverage has several advantages. First, a mobile phone is always close to a network transceiver and therefore needs less transmission power, which saves battery life. Second, channels can be re-used in different cells. The capacity of a network increases as the cells are smaller.

But the price to pay is that cellular networks need to manage the dynamics of subscribers moving between cells, and the problems that brings with it. In this section we'll see how GSM deals with this.

A GSM network consists of three main components: mobile stations (MS), the base station subsystem (BSS), and the network switching subsystem (NSS).

Mobile stations are GSM terminal devices. The BSS manages the radio connections of the cells. The BSS consists of a base station controller (BSC) and base transceiver stations (BTS), which are the actual transmitters and receivers. One BSC can manage various BTS.

The NSS represents the core network part of the GSM network. The key component in the NSS is the mobile switching center (MSC). In essence the MSC is a telephone exchange with extensions to handle the mobility of terminal devices and subscribers.

Subscribers in mobile networks can connect to different MSCs as they move around. Roaming occurs when a subscriber connects to an MSC that belongs to a different network than the home network where he or she is subscribed. The visited location register (VLR) associated with each MSC stores the subscriber data for all subscribers connected to that MSC. VLRs must keep their data updated dynamically as subscribers move around.

Each GSM operator has a database called the home location register (HLR) that holds mobility information for its own subscribers, including to which MSC and VLR a subscriber is currently attached.

Figure 2.22 shows the main components of a GSM network. An MSC is usually connected to more than one BSC, which in turn are connected to

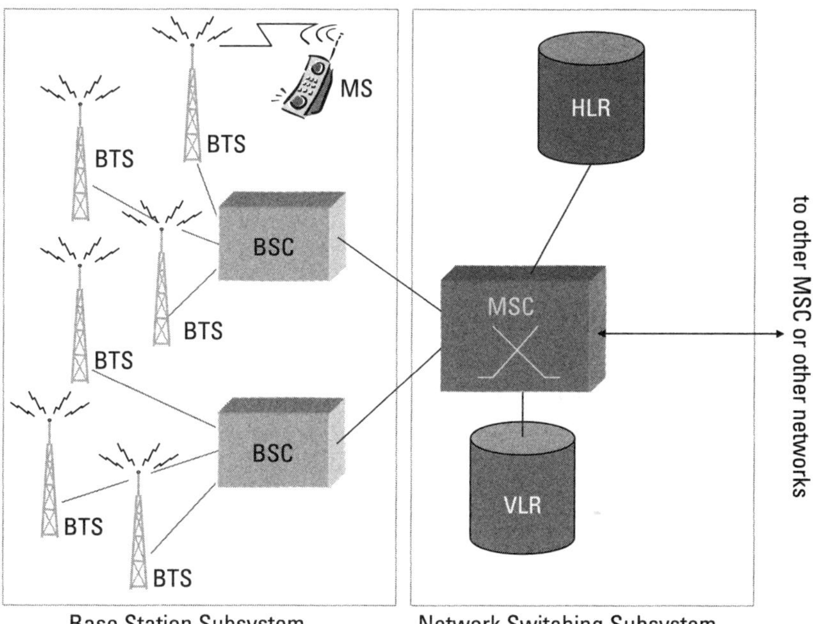

Figure 2.22 Main components of a GSM network.

several BTSs. An MSC can switch connections directly between the mobiles that are connected to it, but it can also forward calls to other MSCs in the same mobile network or switch calls to outside networks. These outside networks can be other GSM networks or fixed networks. An MSC that connected to other networks is usually called a gateway MSC (GMSC).

2.7.2 Mobility Management

The HLR and VLR play an essential role in managing the mobility of subscribers in a GSM network. Since subscribers can move around in cellular networks or even from one network to the other, the network must have some way of locating the subscriber when a call comes in and alerting his or her device.

To solve this problem, GSM networks are divided in location areas. A location area typically covers several radio cells. Each location area has a unique identifier that is transmitted on a special channel in all the cells it contains. Each device monitors this channel. When it detects a change in the broadcast location area identifier, the device knows it has crossed into another location area's radio cell. At that time it requests a location update from the network. There are two ways that a location update can take place:

1. If the new location area is served by the same MSC and VLR, then the VLR registers the move.
2. If the new location area is served by another MSC and VLR, then the mobile subscriber information is moved from the old to the new VLR. The HLR is also updated, so it routes all incoming calls to the new MSC and VLR.

Figure 2.23 illustrates the second case:

1. The mobile device moves into a new cell, notices that the location identifier for this cell is different, and requests a location update.
2. The VLR requests the subscriber information from the HLR.
3. The HLR sends the subscriber information to the VLR and registers that the subscriber is now attached to the new VLR.
4. The HLR informs the old network of the move and orders the old VLR to remove the record for this subscriber.

A device can move from one cell to another while in conversation, and GSM has mechanisms for avoiding the connection and the call to be lost. During a call, a device constantly monitors the quality of the radio signal from

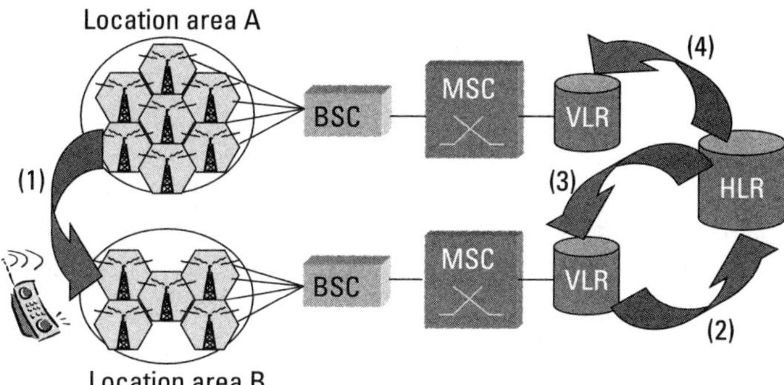

Figure 2.23 Location update.

the current cell, but also from the cells around. Similarly the BSC monitors the radio signals from the mobile devices.

When the terminal device or BSC perceives a decline in the connection quality, the BCS notifies the MSC, which then looks for a better channel in a neighboring cell. The BSC then instructs the mobile device to detach from the radio channel of the old cell and attach to the new channel in the new cell. This procedure of real-time cell change is called *handover* or *handoff* (the latter is mainly used in the U.S.). A handover can trigger a location update, if the new cell is in a different location area.

2.7.3 Mobile Network Identifiers

Another important issue in GSM is security, because radio connections are sensitive to tampering and eavesdropping. Security in GSM involves both authentication and encryption.

A small secure smart card called the *subscriber identity module* (SIM) stores a GSM subscription and becomes associated with a mobile device when inserted. Each subscription is identified by a unique identifier, the international mobile subscriber identity (IMSI). The IMSI consists of a country code, network code, and subscriber identity.

But the IMSI is not the number one dials to call a mobile subscriber. The dialed number, called the *mobile station ISDN number* (MS-ISDN), is different from the IMSI to allow a subscriber to keep his or her number when changing operator, and to allow changes in numbering plan without changing the network internal subscriber identity. The HLR stores the mapping from MS-ISDN to IMSI.

To enhance security, a mobile device communicates its IMSI to the network only when it switches on or does a location update. At this point, the network authenticates the subscriber identified by the SIM in the mobile device, using a secret key algorithm. After authentication the VLR assigns a temporary identifier to the mobile device that is only locally unique, the temporary mobile station identifier (TMSI).

While the mobile device stays connected to the same VLR, it uses the TMSI rather than the IMSI for identification. The TMSI has the advantages of being much shorter than the IMSI and of preventing the IMSI from being sent over the air frequently, which would make GSM more prone to security breaches.

The VLR stores the relationship between IMSI and TMSI, and also keeps track of the location area of the mobile device in the form of the mobile station roaming number (MSRN). The MSRN is an identifier composed of the IMSI and the location area identifier (LAI) of the cell where the mobile device is located. The MSRN therefore contains both the identity and the current location of the mobile device. With every location update the HLR stores the MSRN, so it always knows where the mobile device is located. Figure 2.24 illustrates where the different identifiers are used in the GSM network.

During a call the exchange of digitally encoded voice is encrypted using the same secret key as for authentication, but using an algorithm that is optimized for stream cyphering. But GSM supports more services than voice telephony, with the short message service (SMS) being one of the most prominent ones.

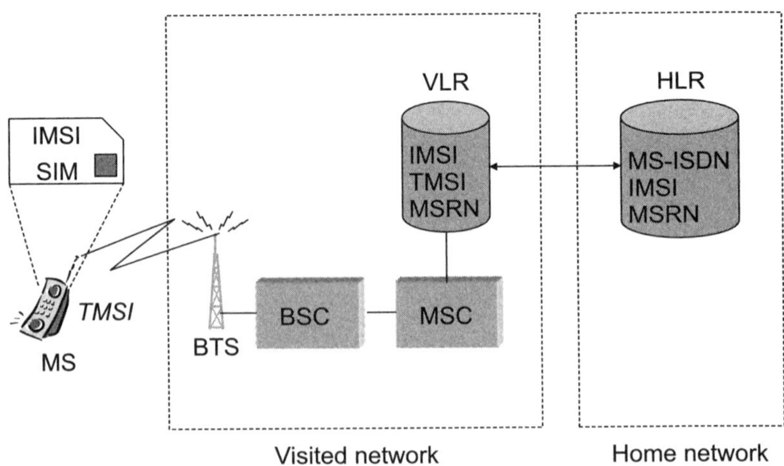

Figure 2.24   The use of identifiers in GSM.

This section should have convinced you that mobile networks have added complexity when compared with fixed telephony networks. Mobile networks can certainly benefit from IN too, but not in a trivial way, so IN had to be adapted to the mobile environment.

2.8 CAMEL and WIN

Being the source of the GSM specifications, ETSI undertook the necessary work to adapt IN to GSM. The resulting ETSI standard is called *customized applications for mobile network enhanced logic* (CAMEL). Putting it in a simplistic way, CAMEL is IN for mobile networks.

Why was this work not done in ITU-T, the originators of the IN standards? In fact it was: IN CS-2 and CS-3 have features to support mobile networks. The BCUP, for example, can be used to let mobility management procedures trigger services. But ETSI had technical and political motivations to come up with their version of IN for mobile networks. On the technical side, ETSI did not find the CS-2 and CS-3 standards sufficiently suitable for GSM networks. On the political side, keep in mind that the telecommunications industry in the 1990s considered GSM a European rather than a worldwide standard. A global standards organization such as ITU-T could not afford to bias its IN standards toward a mobile network technology that was seen to be specific to a certain region.

Table 2.8
CAMEL Phases

CAMEL Phase	Published	Main Features
Phase 1	1996	First version based on IN CS-1, interaction between the SCF and HLR (any time interrogation)
Phase 2	1997	Extended call model, in-band user interactions with SRF, charging features
Phase 3	1999	Triggering on outgoing short messages, triggering on mobility management procedures, triggering on mobile packet data services (GPRS)
Phase 4	2002	Call party handling, triggering on incoming short messages, optimal routing support, triggering on IP multimedia services

ETSI standardized CAMEL in phases. These CAMEL phases resemble the IN capability sets defined by the ITU-T, although they differ both in timing and in scope. Table 2.8 shows the four CAMEL phases and their main features.

In 1998 ETSI and other standards organizations created the 3GPP, which then took responsibility for the CAMEL specifications. ETSI and the other 3GPP members endorse the 3GPP specifications as standards.

Meanwhile in the U.S., a very similar standard was born for the American mobile networks. The Telecommunications Industry Association (TIA) standards committee TR45.2 specified the wireless IN (WIN) standards. The protocol base for this standard is IS-41, the American counterpart of ETSI's mobile application part (MAP) protocol. Although CAMEL and WIN work with different mobile networks and use different protocols, their conceptual models are almost identical. As CAMEL and WIN are so similar, we choose only one and concentrate mostly on CAMEL in this chapter. An excellent reference to wireless intelligent networking that focuses more on the American standards is the work by Christensen et al. [1].

2.8.1   CAMEL at the Distributed Functional and Physical Plane

CAMEL's functional entities are largely the same as that of IN, as demonstrated in Figure 2.25. The SSF handles call and connection control, the SCF executes service logic, and the SRF takes care of special user interactions.

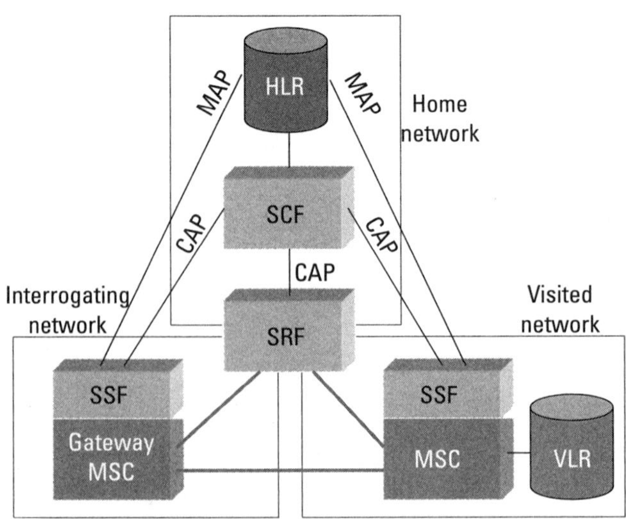

Figure 2.25   CAMEL at the distributed functional plane.

However, CAMEL also differs from IN on important points. In fixed telephony networks, the SSF, SCF, and SRF almost always belong to the same service provider. But as mobile networks allow subscribers to roam into other networks than their home network, CAMEL requires interactions between functional entities in a subscriber's home and visited network.

As in IN, CAMEL executes a service in the home network SCF of a subscriber, but where CAMEL triggers a service depends on whether the subscriber places or receives the call. In the case of an outgoing call, the visited network's MSC triggers the service. In the case of incoming calls, both the GMSC of the home network and the MSC in the visited network can trigger services. This is because GSM switches all incoming calls to a subscriber's home network, regardless of the subscriber's actual location. If the subscriber happens to be roaming, the subscriber's home network GMSC extends the connection toward the visited network, a procedure that is sometimes called *tromboning* in GSM jargon. So an incoming call always involves the home network MSC and may involve a visited network MSC. Figure 2.26 illustrates how CAMEL can trigger services for incoming and outgoing calls.

Newer GSM systems can use optimal routing, which allows calls to be switched in the most efficient way without necessarily passing through the

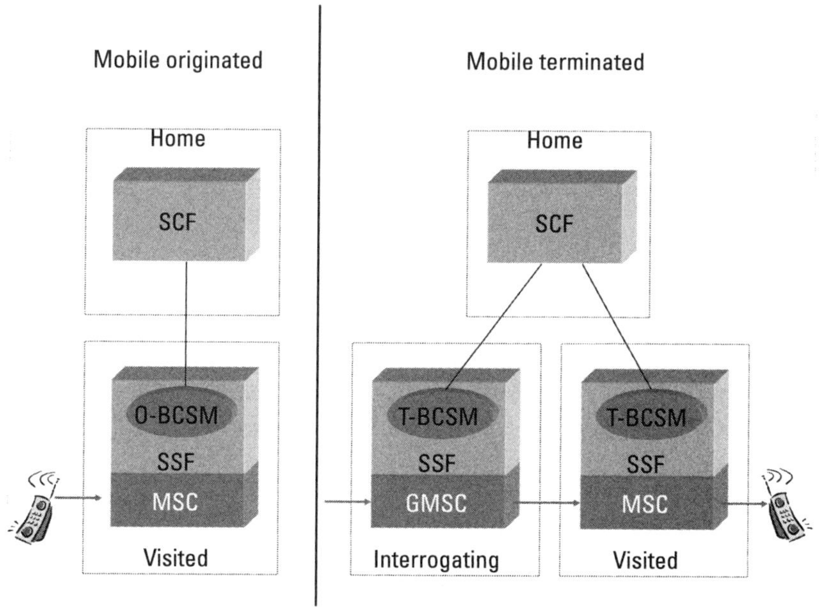

Figure 2.26 CAMEL control for mobile originated and mobile terminated call.

subscriber's home network first.[6] In this case the call may be switched through networks other than a subscriber's home network. CAMEL therefore introduces the more general notion of the interrogating network as the network that can trigger services. In most cases the interrogating network is the home network, but in case of optimal routing it can also be another network that the call passes through.

Another feature that distinguishes GSM from a telephony network is the HLR that stores the subscriber location information. From the IN point of view, the HLR can be seen as a special kind of SDF, and therefore in CAMEL the SCF can communicate with the HLR.

GSM defines precisely when the interactions between MSC and HLR take place (e.g., on power-up of a mobile device, when doing a location update, or during a handover). In CAMEL the SCF can query the HLR at any time to ask about the status of a subscriber, using the standard MAP protocol. This feature is called *anytime interrogation* (ATI) in CAMEL.

CAMEL uses an adapted version of INAP called the CAMEL application part (CAP) for the interactions between SSF and SCF and between SRF and SCF. SS7 carries CAP messages in a similar way as it carries INAP and MAP messages.

The CAMEL basic call state models are almost identical to those of IN shown in Figure 2.14, with a few simplifications:

- CAMEL combines the O_Null_&_Authorize_Origination_Attempt and O_Collect_Info states of the O-BCSM into a single state.
- CAMEL combines the Select_Facility_&_Present_Call and T_Alerting states of the T-BCSM into a single state called Terminating_Call_Handling.
- CAMEL has no O_Mid_Call and T_Mid_Call events.

CAMEL defines additional state models to control GSM-specific services that are not call related, most importantly SMS and the general packet radio service (GPRS). CAMEL phase 4 also provides state models for the Internet multimedia subsystem (IMS) defined as part of the 3GPP third generation (3G) mobile network standards. Figure 2.27 shows how the CAMEL state

---

[6] Although optimal routing for roaming subscribers saves network resources, mobile network operators hardly use it in practice. The standard GSM switching procedure allows an operator to charge for the extension of incoming calls to the visited network of a roaming subscriber. With optimal routing, a subscriber's home network operator loses this income. Although optimal routing could benefit both the network (more efficient use of resources) and the subscriber (less expenses), it threatens a lucrative business for network operators.

model for SMS allows detection points to be set when messages are sent and submitted. As GPRS is a packet data service and IMS provides multimedia services over 3G mobile packet data, we'll save these for the next chapter where we take a closer look at packet networks.

CAMEL employs the same kind of detection points as IN, and trigger processing follows the same procedure.

2.8.2   Triggering Services from a Visited Network

As we have seen in Section 1.2.2, a service provider enables a service by arming a detection point in the originating or terminating basic call state model. Mobile networks introduce complications, because the home service provider of a subscriber usually does not have the privilege to set triggers directly in the SSF of visited networks.

Since the HLR-VLR interaction is well defined in GSM, CAMEL uses the VLR to trigger services. When a user subscribes to a CAMEL service, the HLR sends CAMEL subscription information (CSI) to the VLR of the visited network or the GMSC of the interrogating network, depending on whether the call is incoming or outgoing. The CSI tells the visited network that incoming or outgoing calls for a particular subscriber need CAMEL treatment. When a visiting mobile subscriber places or receives a call, the VLR or GMSC will first check if the home network provided any CSI for this subscriber and checks the criteria. If the criteria are met, this means that the call needs CAMEL treatment and the SSF will start the O-BCSM or T-BCSM.

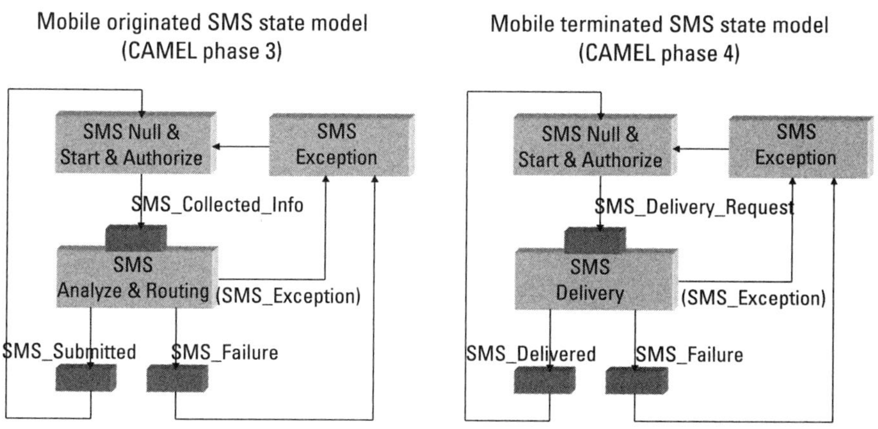

Figure 2.27   CAMEL call-unrelated state model for SMS events.

Table 2.9
CAMEL Subscription Information Types

Subscription Information Type	Shorthand	Held In
Outgoing CSI	O-CSI	Visited network VLR
Terminating CSI	T-CSI	Interrogating network GMSC
Visited terminating CSI	VT-CSI	Visited network VLR

The three most common types of CAMEL subscription information are given in Table 2.9

When a subscriber requires CAMEL processing for outgoing calls, the HLR sends the O-CSI to the VLR. In the case of services triggered for incoming calls, there are two possibilities:

- The service is triggered from the interrogating network: in this case the HLR sends T-CSI to the GMSC of the interrogating network.
- The service is triggered from the visited network: in this case the HLR sends VT-CSI to the VLR.

A combination of both is also possible. The contents of the CAMEL subscription information define the DPs to be armed, the service to be invoked, and the address of the SCF of the subscriber's home network to which the service processing request must be referred. Table 2.10 lists the most important subscription information elements.

Table 2.10
CSI Contents

Item	Description
Trigger detection point list	Identifies at which detection points in the BCSM triggers are set
SCF address	The signaling network (SS7) address of the SCF of the subscriber's home network that the SSF must contact for service processing
Service key	An identification of the service to be invoked in the SCF when an armed trigger detection point is met
Default call handling	Defines how the call should proceed in case of an error in CAMEL processing (e.g., if the home network SCF can't be reached)

Apart from the O-CSI, T-CSI, and VT-CSI, there are a few other types of CAMEL subscription information that are used in special cases. An example is the dialed CSI (D-CSI), which does not activate DPs in the BCSM but triggers directly on special dialed numbers. Other examples are the translation information flag (TIF) and network CSI (N-CSI), which we shall not discuss in further detail here.

2.9 Using IN and CAMEL in Practice

IN came in the 1990s with a dream of hundreds of personalized services enriching the telephony experience for subscribers and creating new revenue for service providers. It didn't happen for several reasons. The creation of new services requires lead times on the order of a year or more because service providers cannot afford to deploy faulty services that can affect network operation or services with security issues. Each service has to come with comprehensive security features and exception handling. Service providers must avoid unwanted interactions between different services as much as possible and resolve them at run time when they do occur.[7] All these requirements make even simple services relatively complex and oblige service providers to test new services exhaustively before deploying them in the network.

Also in the 1990s, the Internet began to upstage telephone networks by offering inexpensive (even free) voice calls and multimedia services. In many ways, the Internet fulfilled the IN dream of rich multimedia services, bypassing the telephone network altogether.

Network and service providers ended up using IN and CAMEL for a reduced set of nonetheless very important services. IN enables important services such as number portability,[8] freephone numbers, and calling cards, and brought to the public telephone network services normally found in private

---

[7] The phenomenon of services having unwanted or unpredictable behavior when interacting with other services is usually referred to as feature interaction. Even simple services can interact (e.g., call forwarding can cause call blocking to be bypassed if, say, subscriber A blocks calls from subscriber B, and B calls C while C forwards its calls to A, then B could end up connected to A). In the 1990s, many research centers dedicated time and effort to devising algorithms to avoid, detect, and resolve feature interactions. The problem, however, remains notoriously difficult to solve even today. Apart from the challenge of modeling complex services and networks to deal with the problem mathematically or algorithmically, it is very difficult to capture user intent and expectation of service behavior. Feature interaction research has come up with only partial results, and today's service providers mostly rely on extensive testing and heuristic approaches to the problem.
[8] Number portability allows a subscriber to keep his or her number when changing domicile or even when changing to another service provider.

branch exchanges (PBXs) such as call forwarding, call blocking, call waiting, and virtual private networks (VPNs).

2.9.1 CAMEL Use Case

CAMEL offers similar services in mobile networks, but the single most compelling service that CAMEL enables is the prepaid mobile subscription. Prepaid subscriptions were responsible for a large portion of the growth of GSM in many countries and still are in emerging markets. As it is such an important service, we'll explain how CAMEL implements it.

Figure 2.28 shows how CAMEL lets a prepaid subscriber make a call from a visited network.

1. When the roaming subscriber tries to make an outgoing call in a visited network, the VLR first checks for the presence of O-CSI.
2. If O-CSI is found, this means that the call requires CAMEL treatment and the SSF starts the O-BCSM.

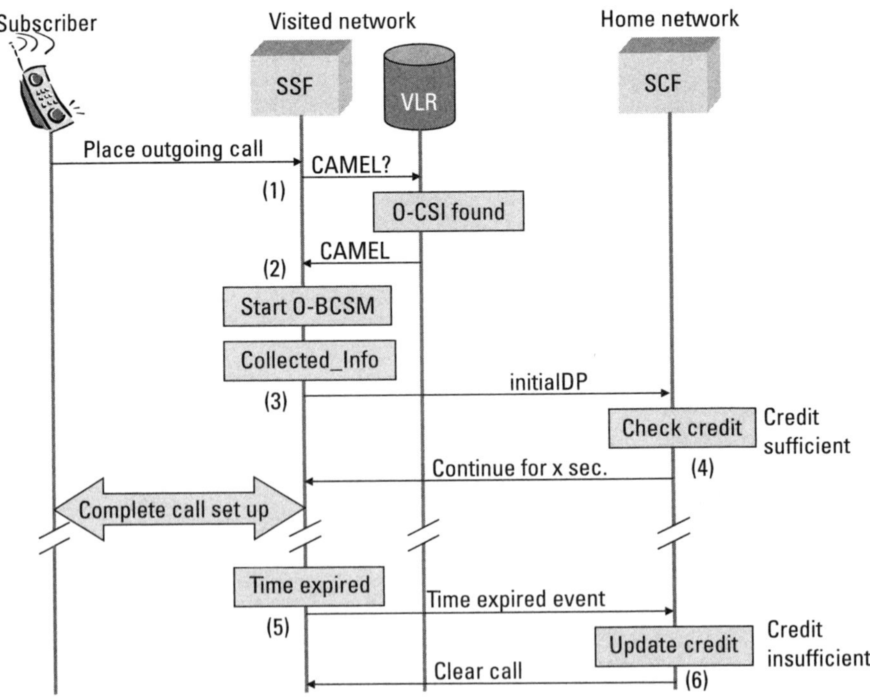

Figure 2.28 Prepaid roaming with CAMEL.

3. The O-CSI contains a trigger for the prepaid service at the Collected_Info detection point. The BCSM encounters this detection point after the user has dialed the destination number and triggers the prepaid service in the SCF.

4. The service logic program for the prepaid service first checks the current credit of the subscriber's prepaid account and instructs the SSF to allow the call to proceed for a limited amount of time. The amount of time that the SCF allows the SSF for the call depends on the visited network location, the destination of the call, and tariff tables for the visited network.

5. The O-BCSM starts a timer and sets up the call. Upon expiration of the allowed time, the SSF sends an new trigger to the SCF. As a result, the SCF updates the credit of the prepaid account and checks again whether the remaining credit is sufficient. If there is sufficient credit to continue the procedure, steps 4 and 5 repeat.

6. If there is insufficient credit as in this example, the SCF clears the call.

This is, of course, a simplified example that shows CAP messages in simplified form. In reality, the user would first hear a message saying that his or her credit has expired before the call is cleared, which would involve interaction with an SRF.

But even in its simplified form, the example does demonstrate that CAMEL provides opportunities for implementing new services that could not be realized with GSM only, like prepaid accounts with roaming. Readers interested in the full details on CAMEL standards and services can refer to the work of Noldus [2].

2.9.2 Commercial IN and CAMEL Offering

Bigger as well as smaller telecommunications vendors offer IN and CAMEL solutions. Random examples include Alcatel-Lucent, Ericsson, OpenCloud, and Teligent Telecom.

Commercial IN and CAMEL products tend to respect the INAP or CAP protocol but only loosely adhere to the SP, GFP, and DFP specifications.

The idea of easy and rapid service creation by chaining together a few SIBs in a graphical design tool turned out to be a mirage. IN vendors needed to add extra SIBs not specified in the IN standards to handle the complexity of real services mentioned earlier in this chapter, and a real service could easily require hundreds of SIBs to cover all use cases, exceptions, and interactions with other features.

Advances in software engineering also challenged the concept of the SIB itself. Computer scientists will notice from Section 1.2 that SIB processing is neither object-oriented nor like remote procedure call (RPC). Objects require encapsulation of data and of operations on that data, while RPC requires processing to resume at the same point where it was suspended. The IN GFP satisfies neither requirement. SIBs are a hybrid and weakly defined computing model, somewhere in between objects and procedures.

As object-oriented design and programming went mainstream during the 1990s, it became a much more attractive model for creating IN services. Most commercial IN platforms use state-of-the-art object-oriented programming tools and design methods.

The Internet provides perhaps the most compelling opportunity to blend telephony with everyday applications such as calendars, electronic mail, and social networks. Ironically, the Internet also challenges the very concept of circuit switched telephony, whether fixed or mobile. Today's IN systems tend to be forward-looking platforms that can control IN services as well as Internet-based multimedia services, allowing service providers to migrate gradually from circuit switched to all packet core networks. The next chapter looks at how the Internet and IP networks in general are changing the telecommunications landscape and how it affects (telecommunications) services and applications.

References

[1]   Christensen, G., R. Duncan, and P. G. Florack, *Wireless Intelligent Networking*, Norwood, MA: Artech House, 2000.

[2]   Noldus, R., *CAMEL: Intelligent Networks for the GSM, GPRS and UMTS Network*, Chichester, UK: John Wiley & Sons, 2006.

3

Value-Added Communications Services in Packet Switched Networks

Telephone networks have ruled the world for more than a century, but, strictly speaking, data communications existed before telephony. Samuel Morse's telegraph created the first data communications network based on electrical signal transmission in the mid-nineteenth century, before Bell invented the telephone.

The proliferation of computers from the 1960s started the evolution of digital data networks that eventually lead to the birth of the Internet protocol (IP), the Internet, and more recently the World Wide Web (WWW). Until about 20 years ago, telephony and data networks used to be two different worlds with different services, different technologies, and different players. Today, we use the Internet not only for data communications, but also for voice and video calls, conferencing, instant messaging, video streaming, and content sharing. The boundaries between telephony and data communications have faded.

IP networks enable many of the services we've discussed in Chapter 2 of this book. Although many operators maintain circuit switched networks for telephony today, IP networks are on their way to becoming the force of convergence for voice and data communications—a good reason to take a plunge into the world of IP networks and the Internet, and see how they are shaping the future of telecommunications.

3.1   IP Networks and the Internet

While everybody has heard of IP networks, the Internet and the web, it's some-times a challenge defining exactly what they are and how they differ. Before going into how these networks can offer voice and multimedia services, let's first straighten out a few basic concepts. An IP network or internet (notice the lowercase i) consists of networks interconnected with the Internet protocol (IP). IP networks can be either publicly accessible or restricted in access. The Internet (notice the uppercase I) is the global interconnection of public and private IP networks.

3.1.1   IP Packets and Routing

Interestingly, IP is not really a protocol, or at least it's not a single protocol. It's more accurate to say that IP denominates a group of protocols for transporting packets of data or datagrams[1] over interconnected data networks.

The Internet Engineering Task Force (IETF) specifies all IP-related protocols (www.ietf.org). Unlike many other organizations, the IETF accepts contributions only from individuals, not companies, and allows anybody to participate without any formal membership or fees. However, the Internet Engineering Steering Group (IESG) and Internet Architecture Board (IAB) that oversee the work of the IETF do put strong requirements on the quality and fairness of the specifications. In practice, you have to earn your peers' respect in the IETF if your contributions are to go anywhere.[2]

First and foremost, IP defines the structure and size of the data packets. Each packet has two parts: a header that contains information about the origin, destination, and contents of the packet, and the body that contains its pay-load (i.e., data). Both the packet header and body can vary in size. The header usually takes up 20 octets (160 bits) but can grow to 160 octets if it contains optional information, which is rarely used. The body can contain anything

---

[1] Although the Internet Engineering Task Force (IETF) documents refer to datagrams, we will use the more common and informal term (IP) *packet* throughout the book.

[2] The IETF is an organization open to participation by anyone and chartered by the Internet Society (ISOC) to specify the standards that enable the Internet. The IETF is structured into working groups belonging to technical areas, and each area director is a member of the Internet Engineering Steering Group (IESG), which sets the overall technical directives. The Internet Architecture Board (IAB) oversees the activities of the IETF and IESG, and arbitrates in case of complaints to ensure that the activities and publications of the IETF are in line with the overall architectural policy of the ISOC. The ISOC has chartered the Internet Assigned Numbers Authority (IANA) to coordinate the assignment of protocol parameters that must be globally unique, such as IP addresses, domain names, and TCP ports.

between 0 and 65536 octets of data. The underlying physical network that carries the packet often puts a limit on the packet size. Ethernet, for example, allows packets to have a maximum size of 1500 octets, header included.

The source and destination of an IP packet can be desktop computers, servers, tablets, smart phones, webcams—indeed any device that can send or receive digital data. To the enable delivery of a packet to its destination, the destination device is identified with an IP address, a 32-bit number. To make them more readable to humans, we usually write IP addresses in dotted decimal notation consisting of four decimal numbers between 0 and 255 separated by dots, where each number represents an octet (8 bits). For example, the IP address of the computer on which I am writing this text is `01011000 00000110 11110000 11001111` and its dotted decimal notation is `88.6.240.207`, which is much easier on the human eye.

Packets are directed through the network by special devices called *routers*. When a router receives a packet, it looks at its header to determine the destination and consults a routing table to decide where to direct the packet next. The next destination for the packet may be the destination device, or it can be the next router on the path to the destination. Figure 3.1 illustrates the principle of IP routing.

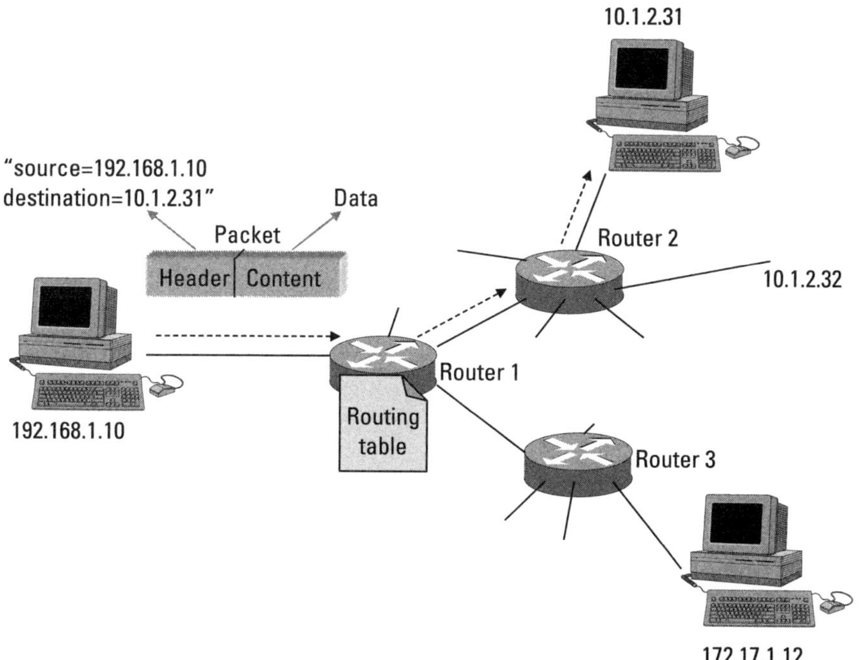

Figure 3.1   Principle of routing IP packets.

Routers use several standardized protocols to create and maintain their routing tables, assign computers with IP addresses, and notify adjacent routers of conditions such as failures and traffic saturation. Although the routing protocols have become more sophisticated and efficient over time, the IP routing principles remain relatively simple. IP routers have autonomous control over how to forward a packet (and whether to forward it at all, as overloaded routers may decide to discard incoming packets) and treat all packets equally and independently regardless of their content. This means among others that packets belonging to the same data flow can follow different routes through the network, arriving with different latencies and in any order regardless of the order they were sent.

3.1.2   IP Protocols

An IP network provides a well-defined but unreliable service for routing packets from source to destination with no guarantee of integrity, latency, or order of delivery. This is why the routing protocols are often used in conjunction with other protocols such as the transmission control protocol (TCP) to improve data transport reliability. These protocols are organized in layers, each taking care of a particular function. Figure 3.2 shows the IP protocol stack.

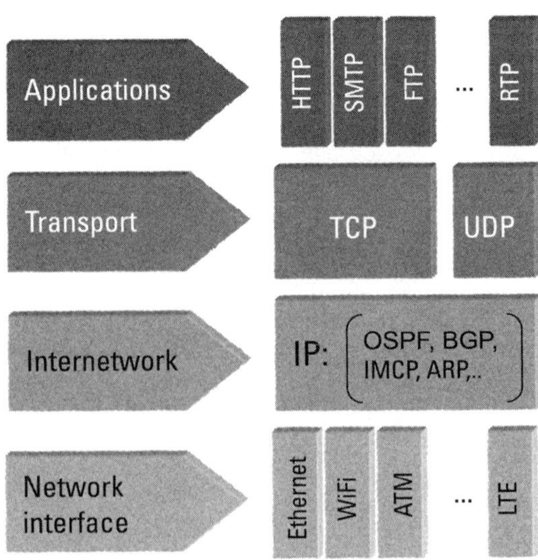

Figure 3.2   The IP protocol stack.

The network interface layer[3] represents the underlying physical network. It can use almost any type of physical channel. The only assumption that the other layers make about this layer is that it can transmit digital data on a point-to-point link. It does not have to be reliable, and it does not have to use packets. Examples of physical networks used to carry IP packets are Ethernet, asynchronous transfer mode (ATM), 802.11 networks (WiFi), or cellular mobile networks.

The internetwork layer provides a concept of a virtual internetwork consisting in reality of many interconnected networks. It defines the IP packet format for sending data through this internetwork. It also defines the protocols to route these packets such as the open shortest path first (OSPF) protocol, the border gateway protocol (BGG), and auxiliary protocols such as ICMP, IGMP, ARP, and RARP, but we shall not discuss them in detail here. Tanenbaum [1] and Stallings [2] explain these protocols in their excellent textbooks that have become classics.

The transport layer provides the end-to-end communication facilities that allow applications to exchange data over the unreliable internetwork provided by the IP routers. The most important transport layer protocol is TCP. Seen from the application, TCP provides a reliable connection for sending a continuous stream of bytes from point to point. A few of the things that TCP does are as follows:

- *Multiplexing:* TCP ensures that different applications on a computer with a single IP address can communicate via IP simultaneously and that the right data gets sent to the right application. TCP defines the concept of logical ports for this and assigns a different port number to each application. In essence, a TCP port is no more than a 16-bit binary number (i.e., a number between 0 and 65,535 in decimal terms) sent with a packet to identify the application that sent the data.
- *Stream data transfer:* to an application it looks as though TCP provides a continuous data connection from one point to another. In fact TCP cuts the data stream up in segments and passes these to the internetwork layer for routing them to their destination as IP packets. At the receiving end, TPC reassembles the data stream from the segments received. It sends a sequence number with each data segment

---

[3] The network interface layer is often subdivided into the link layer and the physical layer. The physical layer transmits bits (digital 0s and 1s) over a physical channel. The link layer organizes these digits into frames (which are like small packets) and employs protocols to minimize the bit error rate on the physical link.

that allows the receiving computer to put segments back in order, no matter what order they arrive in. This is necessary because IP routes each packet independently, so packets do not necessarily arrive in the order they were sent!

- *Reliable delivery:* using a system of acknowledgment and retransmission of lost data packets, TCP ensures that all data arrives safe and sound at the other end.
- *Flow control:* TCP makes sure that the sender doesn't churn out packets faster than the network can handle, and the receiver can receive and process. This can be very important when connecting devices of different processing speeds (e.g., a tablet and a high-end server). It also helps in adjusting the data flow to the varying latency in the network and to network congestion.

Because the protocols at the IP level don't provide these features, they are most often used in combination with TCP. For this reason, many people use the term TCP/IP rather than just IP.

A simpler protocol called the *user datagram protocol* (UDP) offers the same multiplexing as TCP but does not provide reliable delivery or flow control. UDP does not employ packet retransmission and requires less processing, making it faster but also less reliable than TCP. Applications that need high transmission speeds but don't critically depend on reliable transfer of packets tend to use UDP. Examples are voice or video over IP applications, which need high throughput but where the loss or corruption of an occasional packet does not significantly affect the service.

The applications layer contains all the protocols that are specific to applications communicating over IP. There are many application layer protocols, like for e-mail (SMTP, POP3), remote login (Telnet), file transfer (FTP), distributed object computing (SOAP), security (SSL, Ipsec), directory access (LDAP), network management (SNMP), and many, many more. New application protocols are still being defined.

All these protocols exchange messages consisting of a header[4] containing protocol control information and a body containing the payload. In most cases the payload is a message from a protocol one level up in the stack of Figure 3.2.

Figure 3.3 illustrates this principle, which is called *encapsulation.* An e-mail application creates an SMTP message (i.e., your e-mail) and hands it

---

[4] Some protocols use both a header and a trailer. At the network interface layer, the trailer often serves to synchronize the sender and receiver of the message, because physical channels only transmit bits and do not have any concept of packets or frames.

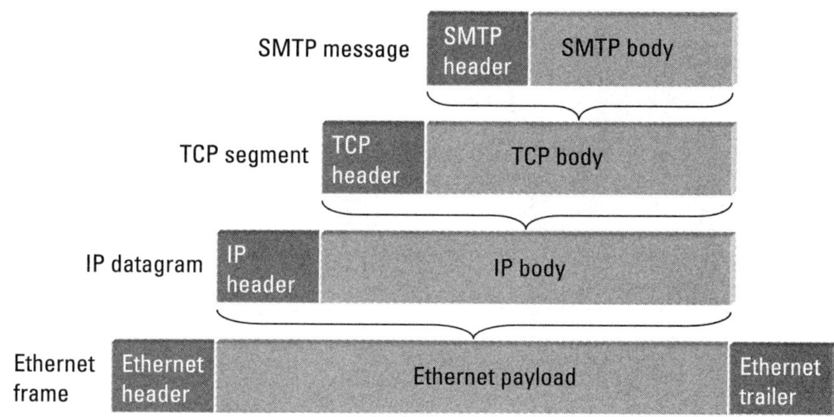

Figure 3.3 Protocol encapsulation.

to a socket, which is the name for the implementation of both the transport layer and network layer on most computers. The socket encapsulates the SMTP message in a TCP segment, which is in turn encapsulated in an IP packet.

The socket then hands the resulting IP packet to the Ethernet driver, which encapsulates it in an Ethernet frame and transmits it on the physical wire. This is how an IP network (using Ethernet for physical transmission) carries e-mail.

3.1.3 HTTP and HTML

One of the application layer protocols that has had the most impact on the evolution of the Internet is the hyper text transfer protocol (HTTP). Scientists at the nuclear research center CERN in Switzerland, and most notably Tim Berners-Lee, defined initial versions of the hypertext mark up language (HTML) and HTTP, the protocol to transport it, around 1991. Since then, HTTP and HTML have made history by creating the WWW.

HTML uses simple in line tags to format and link text. HTTP provides the protocol to request and send HTML documents in plain text format over TCP/IP connections. The browser is the computer application that displays HTML content, enables clicking on links, and fetches and displays hyperlinked documents. One of the first browsers to become popular was Mosaic, introduced in 1993. Mosaic was soon followed by Netscape in 1994 and by Microsoft's Internet Explorer.

HTML has evolved over the years. The recent HTML 5 standard from the World Wide Web Consortium includes features that make HTML more interoperable between applications, more adapted to mobile devices, more suitable for semantic applications, and less dependent on proprietary plug-in applications for video, audio, graphics, and so on. Many developers now use HTML 5 to create web applications that can run on many difference devices, so we will take a closer look at HTML 5 later in this book.

HTTP is a simple protocol that follows the client-server or request-reply model. In a HTTP dialogue, one machine is the requestor (client) and the other is the replier (server). HTTP has only a small number of messages. A client can use a GET message to request data from the server or a POST message to submit data to the server. The server will respond with a message that contains a reply code (200 means ok) and data in plain text format. As with most other protocols we have seen so far, HTTP messages have a simple header and a body that contains the payload data. HTTP uses plain text messages, so any data that is not text (e.g., a picture or video) will have to be encoded in text characters. The multipurpose internet mail extensions (MIME) encoding provides various ways to represent nontext content (e.g., images, video clips, and application data such as PDF files) as plain text and to package several different content items in a single message.

Billions of computers and mobile devices support HTTP because it is the protocol used by web browsers to get data from web servers. Since so many devices already have HTTP installed and since most firewalls will let HTTP traffic pass through, other applications than browsers now also make use of HTTP. For example, developers build distributed object-oriented application with the simple object access protocol (SOAP). Strictly speaking, SOAP is not a protocol at all but a plain text message format for calls to object functions (and their replies) that makes use of HTTP (or any other suitable request reply protocol) for transport. Chapter 5 sheds more light on the use of SOAP.

3.1.4 IPv6

Although 32 bits allow for the addressing of many networks and computers (2^{32}, which is more than 4 billion), the explosion of the Internet has already caused an acute shortage of addresses. A new version of IP, called IPv6, provides a vastly larger address space with 128 bit addresses. To get a rough idea of the vastness of this address space, Andrew Tanenbaum shows in [1] that there are $7{\cdot}10^{23}$ IPv6 addresses for each and every square meter of the earth's surface, oceans included, which is greater than Avogadro's constant (the number of atoms in 12 grams of pure carbon). IPv6 provides enough IP addresses to identify fine grain objects such as individual sensors and nano robots, and perhaps extend the Internet to the moon, Mars, and deep space.

IPv6 uses different, simpler routing procedures than its predecessor IPv4, so as to allow for faster routing. It also provides some standard features that were optional in IPv4 or had to be provided as add-ons, such as encryption, multicasting (sending packets to groups of devices rather than to a single device), and addressing for mobile devices.

The IETF published IPv6 as early as 1996 and its adoption has been reluctant mainly because of the incompatibility between IPv4 and IPv6. IPv6 requires new routers and new drivers that won't connect to IPv4 networks. Although IPv4 has already run out of addresses in some regions [3], network address translation (NAT) makes it possible to expand a single IP address into a range of private IP addresses. If you have a broadband connection at home, it probably uses NAT to provide your desktop, your laptop, your tablet, and your smart phone each with their own private IP address, even though your Internet service provider (ISP) assigned you only a single public IP address for your broadband connection.

NAT can cause problems for applications that rely on specific (public) IP addresses. But despite its problems, networks use NAT on a wide scale, making the need for IPv6 seem less urgent. The deployment of IPv6 gathers more momentum over time, but even so IPv4 will remain in widespread use and will coexist with IPv6 for time to come.

3.2　Mobile Access to IP Networks

As mobile networks and Internet grew up together in the 1990s, an obvious interest arose in accessing the Internet from cellular mobile devices. This did not turn out to be trivial. As we've seen in the previous section, a mobile device needs an IP address to communicate over an IP network, so that the routers in the network can deliver packets to it. When a mobile device moves between network cells, its point of attachment to the network changes. This makes the IP addresses of mobile devices in routing tables volatile and upsets the IP routing mechanism.

By analogy, the postal service needs your home address so that it can deliver letters and parcels to you, and it expects your home address to be durable. If you decide to go live in a camper van and tour the countryside, it becomes difficult for the postman to deliver your mail wherever you may be at any given moment.

Another problem is that second generation cellular mobile networks such as GSM use voice circuits, as we have seen in Chapter 2. A digital voice circuit in GSM provides just enough bandwidth to transport data with a speed of 9.6

Kbps,[5] a mere trickle that provides insufficient bandwidth for most applications. Sending data over circuits is also inefficient because the data generated by applications such as browsing or e-mail does not come as a continuous stream but in bursts. The circuit will waste its bandwidth in between bursts while the device doesn't send or receive data.

It's clear that providing IP access from cellular mobile devices requires a few revisions of the cellular network to allow for efficient packet transmission and to resolve the IP addressing of devices that move around. These revisions were made in the 2.5G, 3G, and 4G standards, the subject of the next section.

3.2.1 From 2G to 2.5G and 3G networks

The 2.5G services such as GPRS enhance the existing 2G network with packet routing facilities that solve the IP addressing problem of mobile devices and provide more efficient data communications. This allows an operator to use most of the existing radio network infrastructure, but puts limits on the network efficiency and the achievable data speed. On the other hand, 3G networks such as UMTS require new infrastructure to improve the bandwidth and efficiency of packet transmissions over the radio link.

ETSI released the first GPRS specifications in 1997 and published a more stable version in 1999. At the same time ETSI also started publishing 3G cellular mobile network standards, but this involved a great deal of litigation and politics because of the high stakes involved. On the commercial side, vendors fought over which patented radio technology to use. On the political side, as ETSI intended to build the 3G standard on GSM and GPRS principles, administrations in other parts of the world labeled these as regional standards, unfit to serve as the starting point for a truly global 3G standard. To defuse this argument, ETSI founded the Third Generation Partnership Project (3GPP) in 1998 together with regional standards organizations CCSA (China), T1 (USA), TTA (Korea), ARIB (Japan), and TTC (Japan). The 3GPP produces technical specifications that its member organizations endorse as standards.

The 3GPP issued the first UMTS specifications in 2000. Since then, the 3GPP also maintains and enhances the GSM specifications and publishes updates of its 2G, 2.5G, 3G, and 4G mobile specifications in periodic releases. The 3GPP publishes a new release more or less every 18 months, as shown in Table 3.1.

---

[5] Kbps stands for kilobit per second, or 1000 bits per second. This unit is also referred to sometimes as kb/s or kbit/s, but in this book we will settle for Kpbs.

Table 3.1
3GPP Releases

Release	Published	Key features
'99	2000	3G radio interface specification
4	2001	Introduces the packet switched (IP) core network
5	2002	Introduces the Internet multimedia subsystem (IMS)
6	2004	Higher speed 3G radio interface (downlink)
7	2007	Higher speed and lower latency 3G radio interface (uplink)
8	2008	4G radio and core network specifications (LTE)
9	2009	Improves 3G-4G interoperability
10	2011	4G specifications that fully comply with the ITU-T definition of 4G in terms of speed and latency (LTE Advanced)
11	2012	Improved interoperation between network operators and service providers
12	2014	Small cells, improved antenna technology for 4G

The 3GPP synchronizes the release of these specifications because 3G standards share many specifications with 2G and 2.5G standards, especially of core network and mobility management functions.

All this did not prevent ANSI from creating its own partnership project, the 3GPP2, because it considered U.S. interests still too weakly represented in the 3GPP. The 3GPP2 specified a system called CDMA2000, which is backwards compatible with IS-95 systems that were widely used in the Americas.

Meanwhile ITU-T set out to mitigate the commercial and political tug of war by specifying a family of 3G network standards, named international mobile telecommunications 2000 (IMT 2000). IMT-2000 family encapsulates several 3G standards, including 3GPP UMTS and 3GPP2 CDMA2000 standards. Other IMT-2000 standards were less successful and never saw commercial deployment.

3.2.2 Mobile Data Networks

As mentioned in the previous section, 2.5G and 3G networks have different radio interfaces but share packet routing facilities in the core network. Figure 3.4 gives an overview of a combined 2.5 and 3G network as many service providers deploy today.

The left half of Figure 3.4 shows the components that provide 2G and 3G radio coverage, the GSM EDGE radio access network (GERAN) and the UMTS terrestrial radio access network (UTRAN). The right half of Figure 3.4 shows the two core networks, a circuit switched network for voice communications and a packet switched networks for carrying data. The elements of the circuit switched cellular mobile network are those we have already seen in Chapter 2: the MSC handles mobile connections and manages subscriber mobility, and the GMSC interconnects to other switched networks.

The structure of the IP packet switched network resembles that of its switched counterpart. The serving GPRS support node (SGSN) provides routing facilities to connected mobile devices and manages mobility, while the gateway GPRS support node (GGSN) manages IP addressing and connects to other IP networks (and in particular the Internet).

3.2.3 The Packet Data Protocol Context

When a device needs data access, the GGSN activates a packet data protocol context (PDP context). A PDP context assigns an IP address to the device and associates this address with the subscriber identity (IMSI) and the network address of the SGSN.

Because the IP address of the device must not change as the device moves between cells, the GGSN sets up an IP tunnel that forwards data destined to

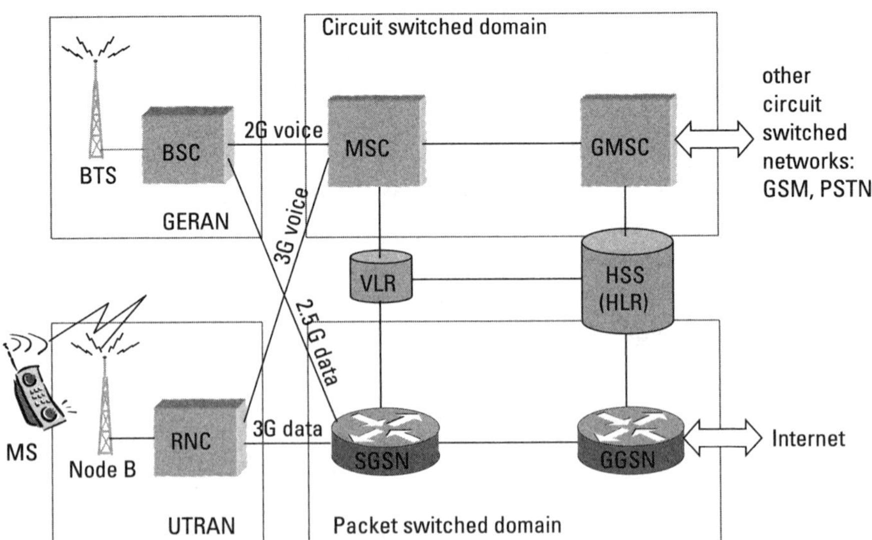

Figure 3.4 Combined 2G-2.5G-3G network overview.

the stable IP address of the device to the actual IP address of the SGSN it is currently attached to. Figure 3.5 illustrates the concept of IP tunneling, which consists of transporting an IP packet in the payload of another. In Figure 3.5, the GGSN has created a PDP context that assigns the IP address x.y.z.q to the subscriber with IMSI i. The PDP context also records that the subscriber is currently attached to the SGSN with IP address a.b.c.d. When the network receives a packet for this subscriber at address x.y.z.q, the GGSN puts this packet in a new IP packet with destination address a.b.c.d, the SGSN to which the subscriber is attached. When the SGSN receives this packet, it extracts the original packet and forwards it over the radio interface to the mobile with IMSI i.

If the subscriber moves to another cell and another SGSN, the GGSN updates the PDP context and IP tunnel, allowing the device's global IP address to remain the same.

3.2.4 Mobility Management in Mobile Data Networks

Just as the 2G circuit switched network is partitioned in location areas to minimize signaling traffic, the 2.5G and 3G packet switched network is divided in routing areas. Like location areas, routing areas embody a compromise between notifying the network of each cell change (which generates lots of signaling traffic even when not receiving data) and paging a subscriber in the entire network when data comes in (which generates lots of traffic when receiving data).

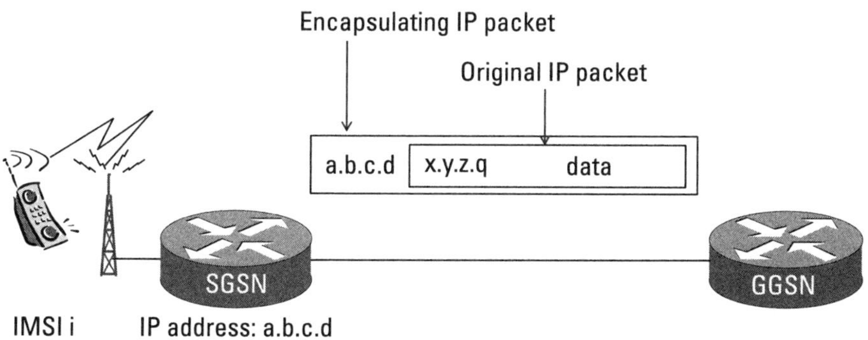

Figure 3.5 Creating a PDP context by tunneling.

In order to facilitate the interworking between 2G, 2.5G, and 3G networks, a location area contains one or more routing areas. In other words, routing areas are a subset of location areas. This has significant advantages:

- 2G location updates automatically imply 2.5G and 3G routing area updates. This avoids the need for duplicate location/routing updates and saves signaling overhead.
- The network can page an incoming 2G voice call in the 2.5G or 3G routing area, which is smaller than the location area. This results in less use of radio resources for paging.

The handover procedure for 2.5G and 3G devices is the same as for 2G, except that it is managed by the SGSN instead of the MSC.

3.2.5 Long-Term Evolution

In its Release 8 (2008), the 3GPP published a new standard for cellular mobile packet data communications that provides significantly higher data rates than its predecessor UMTS and goes by the somewhat odd name long-term evolution (LTE). Although operators and the general public have hailed LTE as the 3GPP's fourth generation (4G) mobile network standard, it doesn't reach the data speeds set by the ITU-T's international mobile telecommunications-advanced (IMT-Advanced) definition of 4G. A later version of LTE, called LTE Advanced (LTE-A), published in Release 10 does satisfy the data speed requirements set by ITU-T: 100 Mbps for high-velocity access (use in cars, trains) and 1 Gbps for low-velocity access (stationary use, pedestrian use).

Most importantly, LTE and LTE-A introduce new radio technology for high-speed packet data communications, but they also introduce changes in the radio and core networks. While they these still resemble the 3G network, LTE has simplified them. Figure 3.6 shows a high-level view of the LTE network. The serving gateway (S-GW) and packet data network gateway (PDN-GW) have roles similar to the SGSN and GGSN in 2.5G and 3G networks. The mobility management entity (MME) handles the mobility procedures we've discussed in Section 3.2.4 such as routing area updates and handover. The HSS provides access to the subscriber database and has the same function as in 3G networks.

What strikes most about the LTE network when compared with the 2G-2.5G-3G network in Figure 3.4 is that it completely abandons circuit switched voice. LTE has an all-IP core network. This raises the question how operators can provide voice communications services with LTE, a question that we'll address in the following sections.

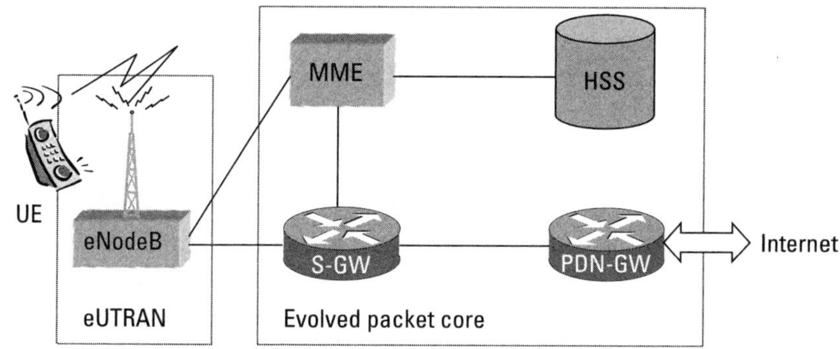

Figure 3.6　LTE network architecture.

3.3　Real-Time Communications over IP

Circuit switched networks are ideal for real-time communications such as voice conversations but tend to be inefficient for data communications. Applications such as file downloads, web surfing, and e-mail don't generate continuous data flows, so it's a waste to keep a connection open at times when neither side sends any data. And most data communications don't require transmission in real time anyway: it doesn't matter when an e-mail, a file, or a web page takes a few seconds more to arrive.

IP was designed with data communications in mind. IP protocols efficiently route data between computer systems without the need to set up circuits in the network, but they were not designed to handle real-time communications.

Despite this, we now routinely use the Internet for many types of real-time multimedia communications, from simple voice calls to multiparty video conferences.

3.3.1　Coding Multimedia into IP Packets

Sending voice or any other real-time media over an IP network requires no more than a codec (short for coder-decoder) that digitizes the analog (voice) signal and sends the resulting samples in a stream of IP packets. Many codecs exist, depending on the quality of the voice or video required and the number of bits per second they produce. Table 3.2 gives some examples of codecs the ITU-T has standardized.

One of the most straightforward codecs is G.711: it samples the voice signal 8,000 times per second (the minimum needed to digitize 4,000-Hz

Table 3.2
Codecs

Standard	Bit Rates	Technology
G.711	64 Kbps	A-law and μ-law pulse code modulation (PCM)
G.726	32 Kbps	Adaptive differential pulse code modulation (ADPCM)
G.728	16 Kbps	Low delay code exited linear prediction (LD-CELP)
G.729	8 Kbps	Conjugate structure–algebraic code exited linear prediction (CS-ACELP)

voice according to the anti-aliasing principle) into 8-bit samples, which gives a stream of 64,000 bits per second. More sophisticated codecs can achieve spectacularly lower speeds as low as 5 Kbps, by taking into account that human speech has a certain pattern to it that gives its waveforms certain predictability. Our speech tends to come with a certain rhythm; it favors certain combinations of sounds and avoids others that are unpronounceable. Codecs can use these properties to send only partial information about the waveforms and synthesize their more predictable components. Speech codecs usually perform worse when presented with audio that is not (only) human voice (e.g., continuous tones, music, or noise).

Sending real-time communications over IP networks doesn't come without a price. Each IP packet that comes out of a codec will have a header of 20 octets,[6] as we have seen in Section 3.1.1. Moreover, a codec has to keep sending a steady pace of packets. It cannot afford to wait too long to fill an IP packet, because this generates latency. For example, it takes an 8-Kbps codec 1.5 seconds to fill a 1500-octet IP packet, added to the time needed to send and route the packet to its destination! This is why codecs have to send the digitized voice in short samples that don't add noticeable delay (e.g., 20 milliseconds). If we use an 8-Kpbs codec with 20-millisecond samples, each IP packet carries only 8000 · 0,02 = 160 bits and has an IP header of 20 bytes (160 bits), which is 100% overhead for IP alone! This shows that using IP networks for real-time communications requires a higher bandwidth than just the bits per second generated by the codec. It also shows that the codec, the sample rate, and the bandwidth have a strong dependency.

---

[6] In fact a packet coming out of a codec will not only have an IP header, but also an UDP header (codecs don't tend to use TCP because of its high latency and jitter), which adds another 8 octets to the count. If the codec uses RTP, the overhead of the RTP header (12 octets or more if RTP options are used) must also be added.

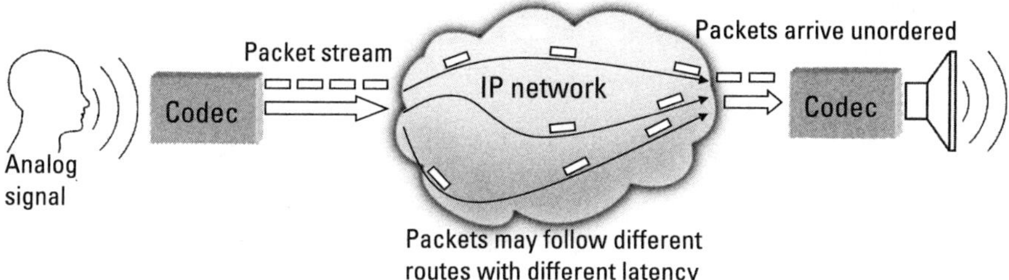

Figure 3.7   Real-time communications over IP networks.

Another problem is that IP networks do not guarantee how long it takes to route a packet to its destination and do not guarantee packet delivery in the same order they were sent. Figure 3.7 illustrates this.

3.3.2   Real-Time Protocol

Packets arriving out of order and with variable latency (also called *jitter*) pose a problem for the codec at the receiving end. Using TCP would solve the packet order but would also worsen the latency and jitter because TCP retransmits lost and corrupted packets. This is why instead of TCP, codecs often use the real-time protocol (RTP) to lend a hand in ordering the packets and reconstructing the time base of the voice or multimedia packets. It would seem logical for RTP to be a transport protocol for real-time communications at the same level as TCP or UDP, but strictly speaking it is an application-level protocol, used with UDP.[7]

Most importantly, RTP adds to each packet a sequence number and a time stamp. The receiving codec uses the sequence number to put incoming packets back in order and the time stamp to correct for jitter. RTP also identifies the media source, so that the codec at the receiving end can relate and synchronize the multiple media streams.

RTP has a companion called the real-time control protocol (RTCP) that allows communication end points to exchange reports about the quality of the RTP stream in terms of packet loss, latency, and jitter. The end points use

---

[7] Since RTP has the rule that an UDP message may not encapsulate more than one RTP message, in practice RTP and UDP tend to melt together. By putting the RTP and UDP headers together, one could consider it a single header of a combined RTP/UDP transport protocol for real-time communications, even if this goes somewhat against the concept of protocol layers providing independent services.

this information to adjust their codec settings (e.g., reducing or increasing the sample size or codec bit rate) or to use a different codec altogether.

3.3.3   Call Signaling for IP-Based Real-Time Communications

From Sections 3.3.1 and 3.3.2, it is clear that IP networks can carry voice, video, and multimedia communications, but it is also clear that circuit switched networks win where it comes to real-time communications efficiency. A telephone network that uses 64-Kbps digital voice circuits can use the full 64 Kbps for voice. No time lost filling IP packets, no header overhead, no jitter caused by packets following different routes. Yet voice over IP is now commonplace, and even telephony networks are increasingly turning away from digital circuits and toward the use of IP. Why? The answer is simply cost.

Circuit switching equipment is expensive because it requires complex protocols (SS7) and must be reliable: a circuit must provide a continuous and consistent data rate; otherwise, the conversation may be compromised or lost. Routers use simpler hardware and protocols, and IP doesn't provide reliable service. While a full-blown core or edge router can cost well in excess of U.S. $100,000, even a mid-size public telephone exchange will cost a multiple of that amount.

The lower cost simply weighs against the shortcomings of IP when used for real-time communications, and some of these can be overcome with brute force. Many service providers just over-dimension their IP network, giving it more than enough bandwidth to mitigate latency, jitter, packet overheads, and packet loss.

Whether they are made over circuit switched or packet switched networks, voice calls and other real-time communications need signaling to alert the called party of an incoming call and to allow the called party to accept or reject the call. Signaling also enables the connecting devices to agree on a codec to use for the communication and other details such as the IP address and TCP ports to use for the voice packet stream. SS7 is not an adequate signaling protocol for real-time communications over packet networks because it negotiates the creation of voice circuits. IP networks use signaling protocols adapted for packet-based voice and multimedia communications. Apart from proprietary protocols such as the ones used by Skype, the session initiation protocol (SIP) and H.323 are the most widely used standard signaling protocols, defined by the IETF and ITU-T, respectively. In the next section we will take a closer look at SIP, which is employed on the Internet and in 3G and 4G mobile networks.

3.3.4   Session Initiation Protocol

The IETF specified SIP as the signaling protocol for negotiating multimedia sessions over IP networks. RFC 3261 [4] published in 2002 defines the SIP version currently in use.

SIP can use addresses of the form `sip:user@host.domain` that the standard domain name service (DNS) of the Internet translates to IP addresses just like it does with e-mail addresses. A example of a SIP address is `sip:johan.zuidweg@ngnresearch.com`. SIP uses plain-text messages and follows a pattern of requests going from client to server and responses coming back from server to client. SIP has a simple request and response format; in fact, plain SIP as defined in RFC 3261 [4] has only six main requests as shown in Table 3.3.

Response messages have a response code like HTTP and are also straightforward. As an example, Table 3.4 shows a few responses that are frequently used in SIP.

Table 3.3
The Main SIP Requests

Request	Meaning
REGISTER	Register location and capabilities of a device to a SIP server
INVITE	Invite a user to a session
ACK	Final acknowledgment that the session is to be established
BYE	Terminate a session
CANCEL	Cancel the setup of a session
OPTIONS	Send and request information about device capabilities

Table 3.4
Example SIP Responses

Response	Meaning
100 Trying	The request is being processed, but the status of this processing is unspecified.
180 Ringing	The destination user has been located and alerting is attempted.
200 Ok	The request has succeeded.

A SIP request has a simple structure. It consists of a request header followed by an optional request body. The following example shows a SIP request header for a call from `johan.zuidweg@ngnresearch.com` to `president@whitehouse.gov`.

```
INVITE sip:president@whitehouse.gov SIP/2.0
CSeq: 11 INVITE
From: sip:johan.zuidweg@ngnresearch.com
To: sip:president@whitehouse.gov
Call-ID: 62439-37@194.140.170.25
Max-Forwards: 70
Via: SIP/2.0/UDP 216.112.6.38
```

The first line is the request line, and it specifies the type of request—in this case, an invitation to set up a call. The subsequent lines are header lines. The `To:` and `From:` lines are obvious.

The `CSeq:` line provides a sequence number for this `INVITE` request, so that the client and server can relate subsequent responses and requests to this transaction. For example, if the user hangs up before the other side answers, the client will cancel call setup by sending a `CANCEL` request with the same `CSeq` number so as to identify the `INVITE` message to cancel.

The `Call-ID:` line provides a unique identifier for this SIP session. Note that a `Call-ID` is different from a `CSeq` because a session can involve more than one request-reply transaction.

Each intermediary server that forwards the request from client to server inserts a `Via:` line to record that it was on the path to the destination. We'll look at the role of intermediary SIP servers later in this section. `Max-Forwards:` limits the number of hops that a SIP request message may make between intermediary servers and avoids cyclic forwarding.

All SIP request messages must contain the header lines shown in the earlier example. In addition, they may contain other header fields such as `Contact:`, `Allow:`, `Accept:`, and `Supported:` that we won't study in detail here and that are well defined in RFC3261 [4].

3.3.5 Session Description Protocol

A SIP request body can contain any type of information, including file attachments, and it can also be empty. SIP typically uses the body to specify the parameters for the multimedia session, using the session description protocol (SDP) format defined by the IETF in RFC4566 [5]. With a SDP payload in the SIP request body, a user agent can communicate information about the media endpoints involved in a session and the media capabilities required.

A SDP session description consists of lines `<type>=<value>` that can specify the session start and stop time, the type of media (voice, video, data), the transport protocol used (RTP, H.320, X.25), media coding (H.261, MPEG), the IP address and UDP port number for the media stream, and even a web address where a more detailed description of the session can be found.

Table 3.5 shows an example SDP description that can be sent in the body of the previous SIP invitation.

SIP can use SDP payloads to negotiate media transfer parameters when setting up a session (in an `INVITE` request), during an active session (using an `UPDATE` request), or even outside the context of a session (using an `OPTIONS` request). This negotiation can take several request-response cycles if either user agent cannot support a parameter proposed by the other (e.g., a voice codec) and has to make a counterproposal.

3.3.6 SIP Dialogues

Up to this point SIP bears certain resemblance to HTTP. It is a request-reply protocol with plain text messages that have a header and a body that can carry a plain text payload. SIP replies have response codes just like HTTP. But HTTP and SIP also differ in an important aspect. HTTP is a stateless protocol: each request-reply transaction is independent of previous or future transactions. In

Table 3.5
Example SDP Session Description

SDP Session Description	Meaning
`v=0`	Version number
`o=Zuidweg 3321578898 IN IP4 194.140.170.11`	Session owner: name, session identifier, and address
`s=SIP and SDP course`	Session name
`i=Teletraining session for SIP and SDP`	Information about the session
`e=johan.zuidweg@ieee.org`	Email address
`c=IN IP4 194.140.170.15/127`	Connection address
`a=recvonly`	Media attribute (receive only)
`m=audio 49170 RTP/AVP 0`	Media description: media name, UDP port number, transport protocol, media format

contrast, SIP has a simple state model and requires that request-reply dialogues follow a certain order.

User agents (UAs), the software that implements SIP in a device, can take on the role of a client (UAC) or a server (UAS) depending on the transaction in course, even within the same session. For example, a UA that receives a session invitation from another UA acts as a server, but if at a later stage it takes the initiative to terminate the session, it acts as a client and the other UA acts as the server. This aspect of SIP differs from HTTP, where clients and servers usually don't change roles. A browser is always a client, and a web server is always a server.

Figure 3.8 shows the basic sequence of SIP messages to set up a session (e.g., a voice call).

1. UA A invites UA B to a multimedia session by sending an `INVITE` message. As we have seen, this message can contain details of the session in SDP format.
2. UA B initially responds with the response code `100 Trying`, thus confirming the receipt of the invite request but not accepting the session yet. While UA B is ringing the user, it can send `180 Ringing` response code to notify UA A of the progress.
3. If the user accepts the call, the UA B sends a `200 OK` response code to inform UA A of the acceptance of the session.

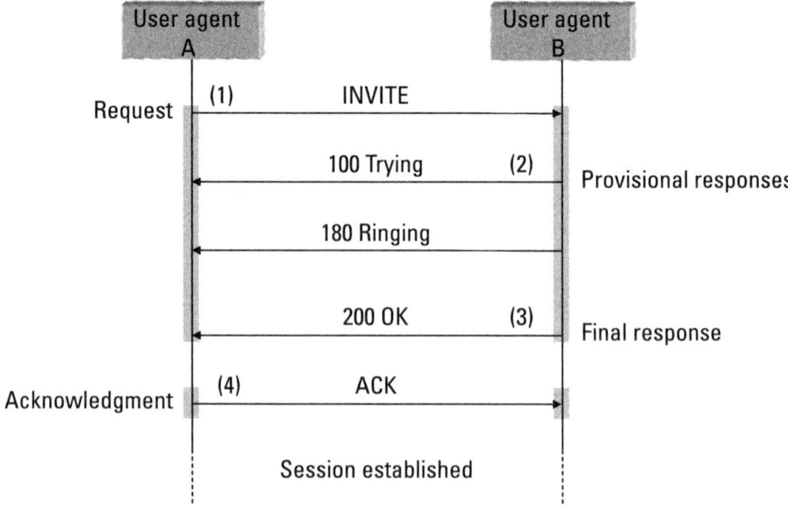

Figure 3.8 Basic SIP session setup.

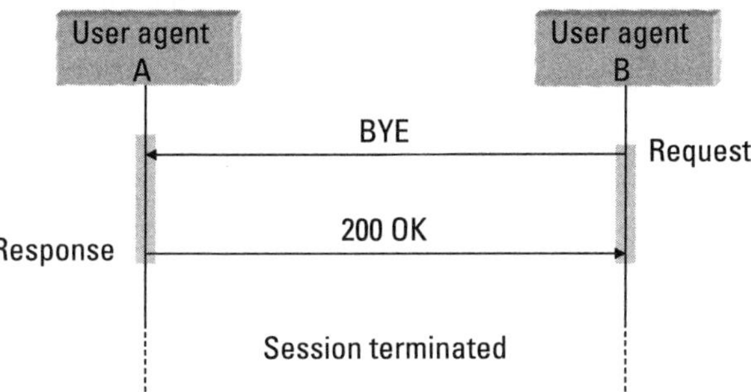

Figure 3.9 SIP session termination request.

4. UA A then completes the three-way handshake by sending an `Ack` message. The `Ack` message is the only SIP request that does not require a reply from the server. At this point, the two UAs will have started the media stream.

In a confirmed and active session, either UA can request termination of the session depending on which side hangs up. Figure 3.9 shows the situation where UA B (which was the server when the session was set up) requests the termination of the session, which reverses the roles of the UAs.

UAs can also cancel outstanding requests (e.g., if a user hangs up when the session request has been ringing for a while without having been picked up by the user on the other side). Figure 3.10 shows this case.

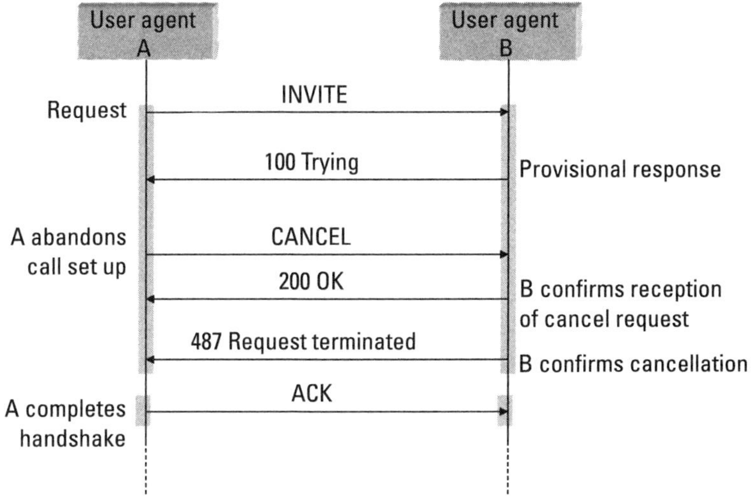

Figure 3.10 Cancelling a SIP request.

For simplicity, Figures 3.8, 3.9, and 3.10 show the most basic SIP message flows and do not show the inside of the SIP messages, which can carry additional session information in SDP format, as we've seen in Section 3.3.5.

3.3.7 SIP Servers

Whereas SIP has the purpose of setting up sessions between UAs, SIP can employ additional types of server:

- *Proxy server:* an intermediate node that processes SIP requests and can perform user location, address translations, security screening, or any other type of processing on the SIP request;
- *Redirect server:* a server that translates the destination address for a SIP request and sends the result back to the client;
- *Registrar:* a server that registers user agents when they are in a network behind a proxy server.

A proxy server takes an incoming request and forwards it to another server. It can do any kind of processing on the request before sending it onward. It can, for example, translate the destination address (call forwarding), refuse the request (call screening), or multicast the request to several new addresses. A proxy server can forward the request to the destination device, to another proxy, or to a redirect server.

What the proxy server does with the requests and where it forwards the request depends on the application that runs on the proxy. A proxy server acts both as a server (to the client that sent the original request) and as a client (to the server to which it forwards the request), so the name proxy server is somewhat misleading. A UA does not need to know about the existence of a proxy server between itself and the other UA.

The redirect server has a more specific function. It translates the destination address of an `INVITE` request and sends the result back to the requesting client (a UAC or a proxy server). The client decides what to do with the result and how to proceed with the call setup. Because the redirect server sends a request back to its origin, the client is always aware of its existence. In this sense the redirect server differs fundamentally from the proxy server. One could compare the proxy server with the pipe mechanism in Unix and the redirect server with a remote procedure call (RPC).

Although SIP can set up sessions directly between UAs, most public and corporate networks will require SIP messages to go through proxy servers for addressing, security, and session processing. If a UA is behind a proxy server, it will need to register to the server to make it aware of its existence, so that the server can forward its SIP messages. This is particularly necessary in mobile networks, where UAs can attach and detach from the network and roam between networks. The registrar is the SIP server that registers UAs. A UA can register by sending a SIP `REGISTER` request to the registrar.

While proxy servers, redirect servers, and registrars have different functions, they often reside on the same machine. It is particularly common to see a registrar and proxy server combined on one physical node. A machine that combines one or more of these server functions (proxy, redirect, or registrar) is often simply called a SIP server.

3.3.8 SIP Servers: An Example

To see how proxy servers and redirect servers cooperate to provide "intelligent" call processing, consider the example of Figure 3.11.

Suppose that A wants to place a call to `sip:helpdesk@ngnresearch`
`.com`. This requires the following steps, shown in Figure 3.11:

1. UA A sends an `INVITE` request to `sip:helpdesk@ngnresearch`
   `.com`. The Internet service provider routes this request through its
   SIP proxy.
2. The SIP proxy looks at the address, determines that it belongs to the
   NGN Research company intranet, and forwards the `INVITE` request
   to `ngnresearch.com`'s SIP proxy.

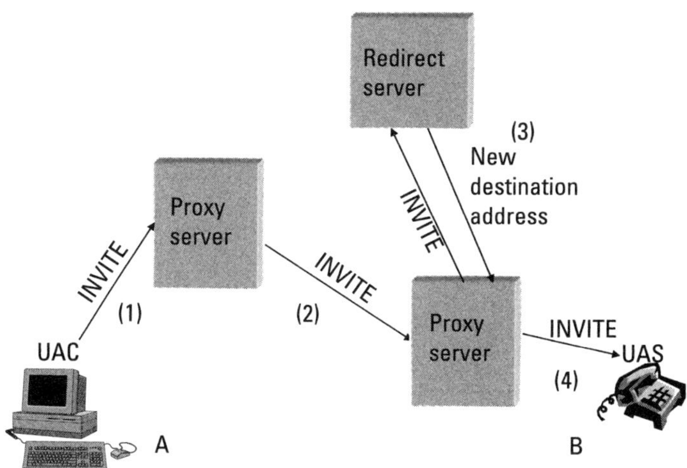

Figure 3.11 Example of a SIP call.

3. The NGN Research SIP proxy first screens the originating and destination addresses for security. It concludes that it is a valid request but detects that `sip:helpdesk@ngnresearch.com` is an alias that should be routed to different people depending on the time of day. It consults the redirect server for the right translation of the address.
4. When the NGN Research SIP proxy gets back the answer from the redirect server, it forwards the request to its destination.

This is a simple example, but it shows the simplicity and flexibility of SIP. Note that the SIP protocol says nothing about the kind of processing that can be done in a proxy server. A call setup can go through any chain of proxy servers that do any kind of call processing. This means that SIP networks are very scalable and flexible.

A SIP proxy can do any of the IN features like those for numbering, routing, charging, access, and restriction we've seen in Section 2.3, but the mechanism of interaction with the call setup process is different. Figure 3.12 compares how a simple service consisting of a call screening followed by a number translation is done in IN and in SIP.

In IN, the SSF controls call setup and delegates control to the SCF for doing the processing. After the SCF has screened and translated a number, it hands back control to the SSF. In contrast, SIP has no centralized call control. SIP processes calls in a distributed way, routing the `INVITE` message from proxy to proxy, where each proxy can do a certain part of the processing. If

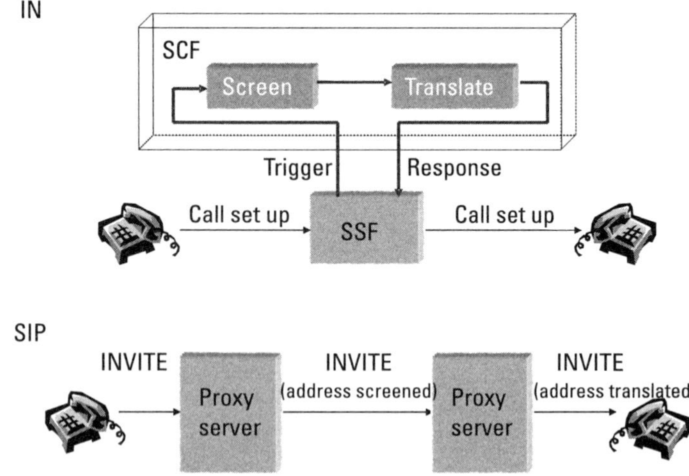

Figure 3.12   IN service processing versus SIP service processing.

you are familiar with the Unix pipe mechanism, this may look familiar. IN, on the contrary, resembles a RPC mechanism.

The most important difference between IN and SIP lies in the tightness of control over the call setup signaling. The SSF in IN keeps tight control of the call state through the BCSM. The SSF can delegate control to the SCF to do service processing, but control always returns to the BCSM.

In SIP there is no central entity that keeps such tight control of the call signaling. A proxy on the path of a SIP message simply sends the request onward after it has done what it has to do.

3.3.9 SIP Extensions

Shortly after specifying SIP, the IETF published several extensions to the protocol, adding a number of new SIP messages. Table 3.6 lists four important extensions that we will refer to later in this chapter and in this book.

The SUBSCRIBE, NOTIFY, and PUBLISH requests introduce a publish-subscribe mechanism for SIP specific events, described in IETF RFCs 3265 [6] and 3903 [7]. With the SUBSCRIBE request, a client can request to be notified of SIP-related events such as a UA changing state from busy to available. A client can use a PUBLISH request to publish an event to a server. The NOTIFY message serves to inform clients of events they subscribed to.

RFC 3265 [6] emphasizes that SUBSCRIBE and NOTIFY can only be used for SIP-related events as defined by the Internet Assigned Numbers Authority (IANA)[2] and IETF, and should not be mistaken for a generic publish-subscribe procedure.

The MESSAGE request, defined in RFC 3428 [8], enables a UA or server to send a free-format text message to another UA or server. Like other SIP messages, the MESSAGE request body carries plain text but may carry MIME-encoded data (i.e., nontext data encoded as text; see Section 3.1.2). This allows

Table 3.6
IP Extensions

Request	Meaning
SUBSCRIBE	Subscribe to events
NOTIFY	Notify a subscribed client of an event
PUBLISH	Publish an event
MESSAGE	Send an instant message

SIP to provide short messaging and instant messaging between UAs and SIP servers.

We will see more of SUBSCRIBE, NOTIFY, PUBLISH, and MESSAGE in Sections 3.4 and 3.5, as IMS and RCS exploit these SIP extensions.

3.3.9.1 Other Protocols for IP Call Signaling

While SIP is widely used, it is by no means the only protocol for IP call signaling. Prior to the publication of SIP, a popular standard for IP multimedia calls was H.323 from the ITU-T. Microsoft's NetMeeting and systems such as Polycom® used this standard for years.

Whereas SIP is a single protocol, H.323 is a suite of specifications describing protocols for registration, authentication, call signaling, and call control, but also codecs, gateways, and voice over non-IP data networks such as X.25. Table 3.7 compares SIP and H.323.

H.323 differs from SIP in that it provides different protocols for call signaling (registration, authentication, authorization, call setup) and call control (negotiation of device capabilities, opening of media channels), all of which SIP offers in a single protocol.

Although H.323 can set up calls directly between two UAs, H.323 networks usually employ a gatekeeper for centralized call control. In this sense H.323 resembles the PSTN and IN, unlike SIP, which works in a more decentralized way.

There have been ardent debates on which is better, SIP or H.323. The truth probably lies somewhere in the middle, but one could say that while H.323 is a widely accepted ITU-T standard and has a broader scope, SIP is simpler, more flexible, and more scalable. It is probably for this reason that SIP now has the upper hand.

Table 3.7
Comparison of H.323 and SIP

	H.323	SIP
Source	ITU-T	IETF
Call signaling	H.225	SIP
Call control	H.245	
Media transfer	RTP	Any protocol
Signaling nodes	Gatekeeper	SIP server
Call management	Tight control, centralized	Loose control, decentralized

Many well-known voice over IP services in the Internet also use their own proprietary protocols. Skype™ is such an example. While these proprietary protocols often resemble H.323 or SIP, they don't interwork with these standards.

3.4 Internet Multimedia Subsystem

In Section 3.2 we've seen that 3G and 4G mobile networks are more optimized for packet switched data traffic than for circuit switched voice traffic. To provide voice communications with these data-oriented networks, an operator has two options. The first option is to fall back to the 2G network for incoming and outgoing voice calls. This requires dual-mode devices that can handle both 2G and 3G (or 4G) communications. Many operators use this technique, but it becomes less of an option as they start phasing out their 2G networks. Some operators have already decommissioned their 2G network because radio spectrum is scarce and expensive, and because maintaining several networks (2G, 3G, 4G) may not be the most cost-effective way to provide cellular mobile communications services.

The other option is to provide voice calls over IP, using the principles we've seen in Section 3.3. Here again operators have two choices. They could concentrate on offering IP data connectivity only and let the subscribers use their favorite third party voice over IP service such as Skype, Google Talk™, Jive™, RingCentral®, Vonage®, and Comcast®. This is called over-the-top (OTT) provisioning of voice services. OTT provisioning of voice communications has several drawbacks for the operator, however.

Although revenue from mobile (voice) telephony decreases over the years, operators still cherish it as a significant source of revenue. This revenue would disappear completely with OTT voice over IP. Another problem is that many voice over IP services use proprietary technology that is incompatible with other providers' services. For example, a Google Talk user cannot connect directly to a Skype user. And voice over IP providers use their own addressing schemes that are not telephone numbers.

In cellular mobile networks, a subscriber anywhere in the world can reach a subscriber on any network anywhere else, using a standardized E.164 number (so called because the ITU-T standardized the international public telecommunication numbering plan in its E.164 series of recommendations). OTT voice over IP services cannot provide such a standard service and experience.

And finally, the relationship of a subscriber with a cellular mobile network operator strongly differs from the relationship that customers have with an Internet-based voice over IP service. A subscriber has a contractual relationship with an operator whereby he or she expects and pays for a guaranteed service. When you pick up the phone you expect a dial tone, and when you dial you expect to get connected (if the other side picks up the phone, that is). OTT services are often free and without guarantees. Even when they're not free, they often come with disclaimers with respect to availability and quality of service. Here again, OTT voice over IP services are disruptive compared with the good old mobile telephony service.

Cellular mobile network operators clearly have an incentive to provide their own voice service over the packet network. Enter the Internet multimedia subsystem (IMS).

3.4.1　IMS Overview

The 3GPP published the first specifications of the IMS in 2002 as part of Release 5 (see Table 3.1 in Section 3.2.1), with the intention of providing a SIP-based infrastructure for providing voice and multimedia calls over 3G mobile packet networks. Since 2002, IMS has evolved with each new 3GPP release.

Figure 3.13 provides an overview of the main IMS functions and their interfaces. To most of us, the IMS specifications look complex and intimidating at first sight. One of the reasons for this complexity is that IMS has to manage the identity and mobility of subscribers. Subscribers of cellular mobile networks move around the network and can roam to other IMS networks, so their point of access can change. This means it requires some administration to route SIP calls to them. In addition, IMS must be able to translate

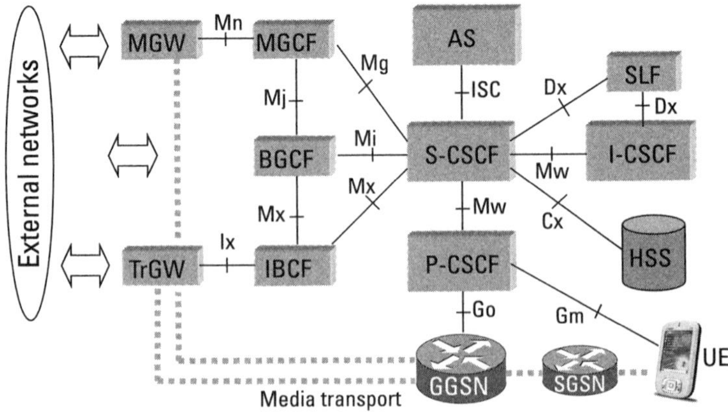

Figure 3.13　Internet multimedia subsystem: overview.

between E.164 telephone numbers and SIP identifiers of the type we saw in Section 3.3.2.

The basics of IMS aren't that difficult to grasp, however, especially if one is familiar with SIP and GSM mobility management, both of which we've seen earlier in this book. Remember that from the SIP point of view, intermediate servers are SIP proxies, registrars, and redirect servers, and they work as we've seen in Section 3.3.3. IMS gives them elaborate names such as S-CSCF, P-CSCF, I-CSCF, and AS, but they're still SIP servers.

Taking a global view one could say that the right-hand side of Figure 3.13 showing the S-CSCS, P-CSCF, I-CSCF, HSS, and AS forms the core of the IMS. The left-hand side showing the MGW, MGCF, IBCF, TrGW, and BGCF contains the functions for connecting to other voice and media networks, not necessarily with IMS. While key interfaces such as Mx and ISC use SIP, some such as Cx use Diameter[8] and other protocols.

3.4.2 IMS Core

We now take a closer look at the core IMS functions and interfaces, and how they interact to set up and control voice and multimedia calls over IP networks. We won't go into too much technical detail, because a complete treatment of IMS would require a complete book and there are good books on IMS available such as [9]. Our purpose is to understand enough of IMS to know how to create services and applications with it.

Subscribers register to the IMS network through the serving call state control function (S-CSCF) in their home network. The S-CSCF takes care of a number of basic functions:

- Registering a subscriber;
- Associating the subscriber's E-164 number with a SIP identifier;
- Routing incoming SIP messages to the subscriber and outgoing SIP messages from the subscriber;
- Invoking value-added services for a subscriber's incoming and outgoing SIP sessions;
- Generating charging records for SIP sessions.

---

[8] Diameter is an authentication, authorization, and accounting (AAA) protocol defined by the IETF in RFC 6733 [19]. A diameter message carries information in the form of attribute-value pairs. Unlike many other application layer protocols, Diameter is a binary protocol and does not send information as plain text. Diameter attributes and values are numerical codes.

From this list it follows that the S-CSCF combines the functions of a SIP registrar and SIP proxy, and its role resembles[9] that of the MSC in GSM and the SSF in CAMEL.

The P-CSCF is the SIP proxy that forwards a subscriber's SIP messages in the network the subscriber happens to be roaming in. Its role bears resemblance to that of the visited MSC and VLR in GSM.

The I-CSCF receives incoming SIP messages for which the IMS network does not know the destination S-CSCF and looks up the S-CSCF of the subscriber to which the message is addressed.

IMS has access to the home subscriber system (HSS), which holds subscriber data and can be regarded an evolution of the HLR. The HSS provides access to a subscriber's identities, authentication keys, S-CSCF address, and may also hold specific data for value-added services and applications.

Application servers (ASs) handle additional service processing, and their role is similar to that of the SCF in CAMEL and IN, although the invocation mechanism differs. ASs play a central role in the creation and provisioning of value-added IMS services, and we'll look at them in more detail later.

To understand how the S-CSCF, P-CSCF, I-CSCF interact to establish a SIP session between two subscribers, consider the scenario in Figure 3.14. This figure shows the situation where the subscriber who originates and the subscriber who terminates the session both roam in networks different from their home network. For simplicity, Figure 3.14 only traces the path of the SIP INVITE request and does not show gateway nodes at the network edges.

The SIP INVITE request from subscriber A to subscriber B may follow this path:

1. Subscriber A wants to make a call to subscriber B. The UAC of subscriber A (i.e., his or her device) sends a SIP INVITE request to the P-CSCF it is registered to in the visited network.
2. The P-CSCF forwards the request to the subscriber's S-CSCF in the home network.
3. The S-CSCF may invoke value-added service logic (e.g., call screening) for this subscriber by sending the INVITE request through an AS for additional processing.
4. Assuming that the S-CSCF of subscriber A does not know the S-CSCF of subscriber B, it sends the INVITE request to the I-CSCF in subscriber B's home network.

---

[9] The S-CSCF resembles the MSC and SSF only to some degree; there are also some important differences we'll discuss further on. We've made the comparison here for illustrative purposes.

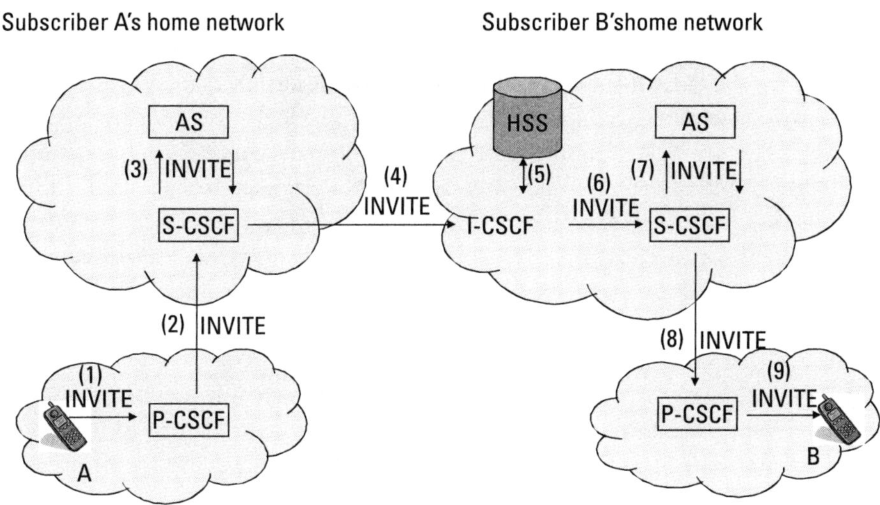

Figure 3.14 Example of session establishment in IMS.

5. The I-CSCF queries the HSS to determine the S-CSCF of subscriber B.
6. The I-CSCF forwards the `INVITE` request to the S-CSCF of subscriber B.
7. The S-CSCF may invoke value-added service logic for subscriber B by sending the `INVITE` request through an AS for additional processing.
8. The S-CSCF routes the SIP `INVITE` request to the P-CSCF serving subscriber B in the visited network.
9. The P-CSCF forwards the SIP `INVITE` request to the UAS of subscriber B.

The SIP response from subscriber B then follows the inverse path. As we've seen in Figure 3.14, servers on the path may return temporary responses such as `100 TRYING` while the intermediate servers in the chain process and forward the request.

IMS has several functions for interconnecting with other networks, IMS and non-IMS. These include the IBCF, TrGW, BGCF, MGCF, and MGW. Their role is to translate between signaling protocols (MGCF, IBCF) and between codecs (MGW) to provide security (BGW, IBCF) and to hide network internal parameters such as media endpoints and network addresses (TrGW). While they have important functions in IMS, we won't describe them in further detail here.

3.4.3 Application Servers

While the S-CSCF, P-CSCF, and I-CSCF provide basic session establishment, application servers (ASs) provide value-added services. The AS runs applications to control SIP sessions, just like an SCF runs service logic to control calls in CAMEL and IN. But whereas the IN standard defines the building blocks that make up service logic, IMS leaves it up to the service provider to determine how to program IMS applications. The only real requirement for an AS is that SIP messages go in and SIP messages come out.

An AS can work in various modes, as Figure 3.15 illustrates.

- *SIP proxy:* the AS processes and forwards incoming SIP messages. When used this way, the AS can screen, forward, divert, or otherwise manipulate SIP sessions just like any SIP proxy. The incoming and outgoing SIP messages will have the same `Call-ID`.
- *Terminating UA:* the AS acts as a UAS terminating a SIP message chain. When used this way, the AS can handle calls to a voice mailbox or televoting service, for example.
- *Originating UA:* the AS acts as a UAC originating a SIP message chain. When used this way, the AS can initiate a wakeup call, for example.

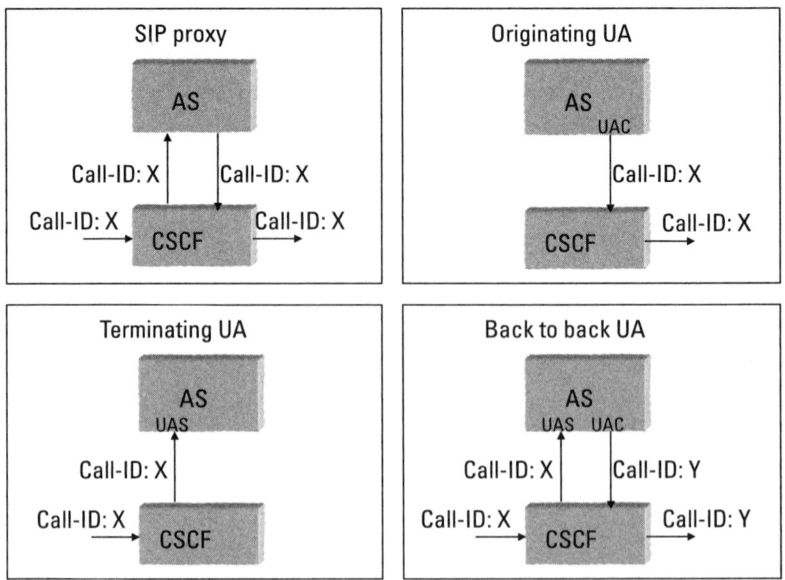

Figure 3.15 AS operation modes.

- *Back-to-back UA:* the AS acts as a UAS for incoming SIP messages and originates a SIP dialog as UAC. When used this way, the AS maintains two SIP sessions, so that the incoming and outgoing SIP messages have a different `Call-ID`. In this configuration, an AS has more control over a session than as a proxy and can set up third-party calls or control services such as collect calling or call when available.

To understand how an AS enables value-added services, we'll look at the example of a call when available service. If a subscriber tries to reach a number that is busy, this service directs the calling subscriber to a web page where he or she can indicate what to do: hang up, leave a message, or call again when the destination becomes available.

In this example, the AS acts as a back-to-back UA and also as HTTP server.[10] Figure 3.16 shows a simplified SIP message flow for this service (omitting temporary responses, confirmations, and ACKs as well as message details for simplicity):

1. Subscriber A sends an `INVITE` to subscriber B. This `INVITE` message traverses subscriber B's S-CSCF and the AS connected to it.
2. Subscriber B is occupied, so his or her UA returns a `486 Busy` response to the S-CSCF, which forwards it back to the AS.
3. The AS activates the call back when available service, and instructs subscriber A's UA to redirect to the HTTP server hosted on the AS. This `302 Redirect` message traverses the S-CSCF and other intermediate servers on the reverse path to reach A.
4. A's UA sends a `HTTP GET` message to the URL of the AS provided in the redirect message. The HTTP server sends back a `200 OK` response that contains a HTML page that displays options from which subscriber A can choose: abort the call, leave a message, call back when available, and possibly other options. The browser on subscriber A's device renders the HTML page.
5. The subscriber clicks on the option to call back when available, and the browser posts the response to the HTTP server on the AS.
6. The AS sends a SIP `SUBSCRIBE` request to subscriber B's UA, asking it to provide a notification when B releases its current call and becomes available.

---

[10] Services that mix real-time communications and Internet-based applications in this way are called *mash-up applications*. Since there are no assumptions on how an AS works on the inside, it can run any mash-up application and host a HTTP server.

7. When B hangs up his or her call and becomes available, his or her UA sends a `NOTIFY` message to the AS.

8. The AS now sends `INVITE` messages to both subscriber A and subscriber B to establish a call between the two.

Step 8 of Figure 3.16 demonstrates how the AS works as a back-to-back UA, representing two UACs that mediate a session between subscriber A and subscriber B. This creates two different SIP sessions with different `Call-IDs`, even if they set up a single media stream between A and B. Remember that we're only talking about signaling here: the media stream between A and B has its own route and does not traverse the S-CSCF and AS!

What follows after step 8 would of course have to be carefully programmed, so as to establish a media stream between A and B only when both A and B respond with `200 OK`. If either of the two rejects the session, the AS would have to cancel the other outstanding `INVITE` request. The service may also need to play or display some messages to either subscriber if there is a failure in call set up. This requires some design and programming effort but is otherwise straightforward.

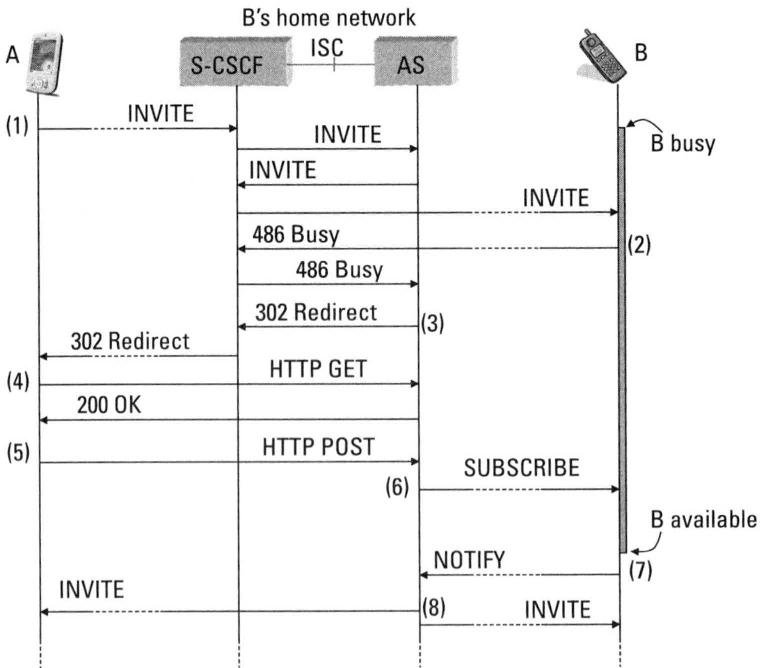

Figure 3.16 Call when available service, example message flow.

3.4.4 CSCF and AS State Models

To facilitate the creation of service logic, IMS defines state models for the incoming and outgoing SIP dialogues at the ISC interface between the S-CSCF and AS. Figure 3.17 shows these state models:

- *Incoming leg state model* (ILSM): maintains the dialogs of incoming SIP messages and decides which messages to forward to the AS. The ILSM effectively triggers services to be handled by the AS.
- *Outgoing leg state model* (OLSM): maintains the dialogs of outgoing SIP messages.
- *Incoming leg control model* (ILCM): handles the SIP dialogue of messages forwarded to the AS.
- *Outgoing leg control model* (OLCM): handles the SIP dialogue of messages received from the AS.
- *AS-ILCM:* handles the SIP dialogue of messages received from the S-CSCF.
- *AS-OLCM:* handles the SIP dialogue of messages sent to the S-CSCF.

When the AS works as a back-to-back UA, the ILCM and AS-ILCM maintain a distinctive SIP session from the SIP session managed by the AS-OLCM and OLCM. The `Call-IDs` for these sessions are different, as we've seen in Figure 3.15.

3.4.5 Service Triggering and Service Chaining

The state models in IMS resemble the I-BCSM and O-BCSM in IN that we've seen in Section 2.5.2. Likewise, IMS triggers services in a way that has

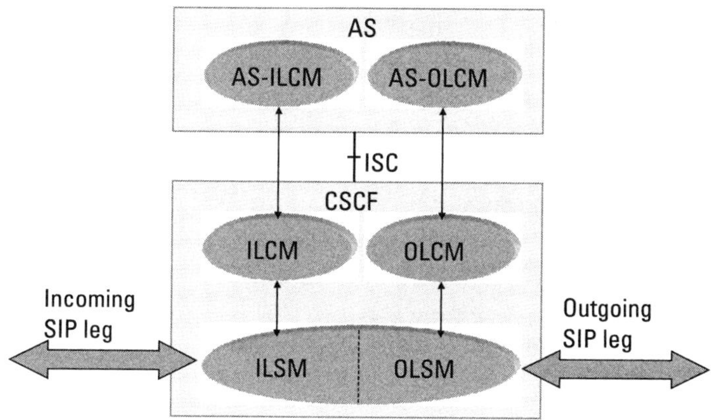

Figure 3.17 State models at the ISC interface between S-CSCF and AS.

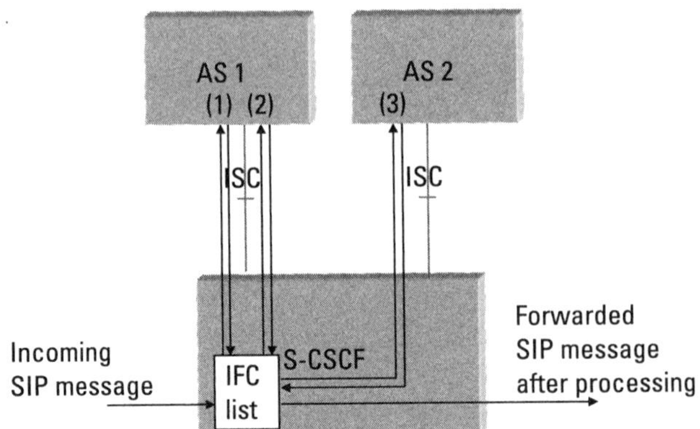

Figure 3.18 Service chaining on conditions specified in the IFC.

similarity with the service-triggering mechanism in CAMEL in Section 2.8.2. For each subscriber, the HSS holds initial filter criteria (IFC) that describe under which conditions to forward a SIP message to an AS for service processing and which AS to forward the SIP message to. The S-CSCF caches the IFC for all subscribers registered to it,[11] much like the VLR caches the CSI in CAMEL (see Section 2.8.2). IFC can specify any Boolean combination of service point triggers (SPT) that include:

- The initial SIP request (e.g., `INVITE`, `REGISTER`);
- The presence of specific headers in the SIP request;
- The content of specific headers in the SIP request;
- The identity of the originating or terminating subscriber;
- Session description information as specified in the SDP payload of the SIP message (e.g., the media format).

Figure 3.18 illustrates how the CSCF chains service triggering, based on the conditions in the IFC. It shows how a subscriber may have a list of several IFC that may invoke the same AS sequentially, but also different AS.

IFC can specify only a single AS to handle service processing and cannot identify particular services to be triggered in the AS. While this reduces the processing burden on the S-CSCF (which has to handle a lot of throughput), it also limits the service-triggering mechanism. The AS may need to have additional logic to trigger specific services when a session needs to invoke complex

---

[11] A S-CSCF can also trigger services for subscribers that are not registered to it. IMS defines unregistered IFC for this purpose, which we won't describe in further detail.

services composed from more elementary ones or when a subscriber has subscribed to different services to be triggered under different conditions. Figure 3.19 shows how the S-CSCF can leave service chaining to the AS. In this case, the S-CSCF passes control to the AS once, even if the session needs to trigger multiple services, and the AS takes care of the detailed evaluation of trigger criteria and service chaining. Note that IMS does not specify any particular mechanism for this. It is up to the vendor or service provider to implement this service chaining functionality in the AS. Figure 3.19 also illustrates that IMS allows an AS to forward a SIP request directly to another AS over the ISC interface, allowing the SIP message to pass through any chain of AS.

The S-CSCF does not handle desired or undesired interactions between services. IMS leaves this up to the AS. Managing interaction between services, especially unwanted feature interaction, is a serious issue and is a complex subject on its own. To get an idea of the scope and complexity of this problem see [10], the proceedings of the latest conference in a decade of conferences on this topic.

The 3GPP has introduced the concept of the service capability interaction manager (SCIM) but does not specify its functionality. The 3GPP stage 2 specification of IMS (TS 23.228, see [11]) merely mentions that an AS may include SCIM functionality without going into further detail.

In practice, vendors provide proprietary SCIM solutions as part of their IMS products. One such example is the Alcatel-Lucet Service Broker™ [12]. Service interaction management across vendor platforms remains a challenge, however.

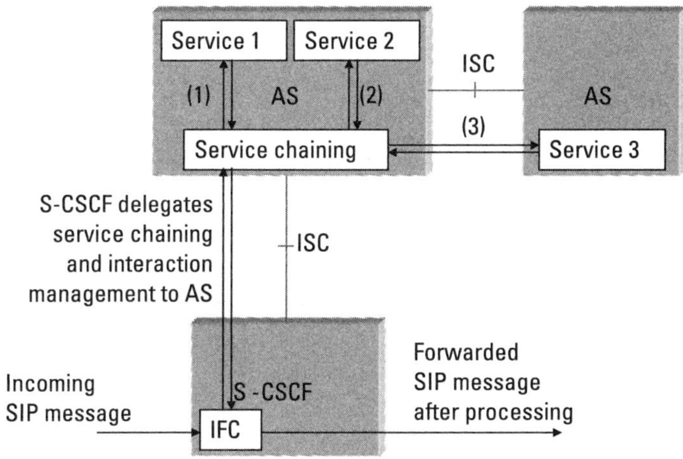

Figure 3.19   Service chaining managed by the AS.

3.4.6   Programming IMS Application Servers

The IMS specifications define how the S-CSCF triggers services and forwards SIP requests to AS for service processing, but not how the AS implements and executes these services. IMS makes no assumptions as to what goes on inside an AS, except that it has an AS-ILCM and AS-OLCM that can receive and send SIP messages.

One can program the service logic in an AS in any programming language and on any platform. In Chapters 4 and 5, we'll take a detailed look at some of the ways to program services in an AS:

- *SIP servlets:* provide a lightweight way to process SIP requests just like HTTP servlets process HTTP requests using the Java programming language.
- *Java APIs for integrated networks service logic execution environment* (JAIN SLEE): provides a more complete Java framework for controlling SIP-based communications services, with extensive back-end processing and integration with databases.
- *Application programming interfaces* (APIs): the AS can also delegate control to external servers through APIs. This is a way to enable third-party control of IMS services.

One of the simplest ways to instruct an AS what to do with a SIP session is through the call processing language (CPL). The next section describes how CPL works.

3.4.7   Call Processing Language

CPL provides an extensible mark up language (XML)[12] format to define call processing scripts. CPL allows for only if-then style decisions; it does not enable real programming, as it does not provide variables or looping.

CPL offers the following constructs, called actions:

---

[12] The extensible mark up language (XML) provides a plain text format to structure information. It structures a document with information elements delimited by an opening tag of the form `<tag-name>` and a matching closing tag of the form `</tag-name>`. XML has only a few simple rules: an XML document must start with the tag `<?xml>`, it must have exactly one root tag, an opening tag must have a matching closing tag, and tags must be nested correctly (i.e., `<tag1><tag2></tag2></tag1>` is correct but `<tag1><tag2></tag1></tag2>` is incorrect). XML tags are case sensitive. XML schema and document type definitions (DTDs) give additional structure to XML documents above the basic XML syntax described earlier.

- *Top level actions:* at the top level, a CPL script describes either an incoming or outgoing action.
- *Switches:* provide decision structures, comparable to the case statement in programming languages. CPL allows switching on address, time, language, and priority of the incoming SIP message.
- *Location modifiers:* modify the destination address, for proxying or redirecting a SIP session. This action can modify the address explicitly or refer to a database for lookup.
- *Signaling actions:* define the SIP message to be sent. The proxy action forwards the SIP request to its destination and allows for conditional action depending on the value of the SIP response returned. Other possible actions are redirect and reject.
- *Nonsignaling actions:* CPL can also define actions that do not generate outgoing SIP messages, such as sending an e-mail or logging session data.
- *Subactions:* enable the definition of complex actions from elementary actions, but without recursion. Subactions resemble subroutines or procedures in programming languages, except they can't make recursive calls.

Table 3.8 illustrates the format of much-used CPL actions. RFC3880 [13] defines the full range of actions and subactions.

Table 3.8
CPL Actions: Examples

Action Type	Example

Switch	`<address-switch` fields to be matched`>`
	   `<address>` action to be taken if address matches
	   `</address>`
	   `<otherwise>`
	   default action if address doesn't match
	   `</otherwise>`
	`</address-switch>`

	`<time-switch>`
	   `<time` parameters specifying the time frame `>`
	   actions to take in the specified time frame
	   `</time>`
	   `</time-switch>`

Table 3.8
CPL Actions: Examples *(Cont.)*

Action Type	Example

Location modifier	`<location url="new destination"></location>`
	
	`<lookup>`
	　`<success>`
	　actions to be taken on successful lookup
	　`</success>`
	　`<notfound>`
	　actions to be taken when lookup did not produce an address
	　`</notfound>`
	　`<failure>`
	　actions to be taken when lookup failed
	　`</failure>`
	`</lookup>`

Signaling action	`<proxy timeout="time in seconds to wait for reply">`
	　`<busy>` actions to be taken on busy `</busy>`
	　`<noanswer>`
	　actions to be taken on no answer
	　`</noanswer>`
	　`<failure>` actions to be taken on failure `</failure>`
	　`<redirection>`
	　actions to be taken when the request was redirected
	　`</redirection>`
	　`<default>` default action `</default>`
	`</proxy>`

	`<redirect />`
	`<reject status="rejection status, e.g. busy">`
Nonsignaling action	`<mail url= "mailto:han@ngnresearch.com?subject=Missed%20call" />`
	`<log name="log file to store the comment in" comment="text to log" />`

Subaction definition	`<subaction id="subaction name">`
	sub-action definition
	`</subaction>`

Subaction reference	`<sub ref="name of subaction to call" />`

CPL does not enable real programming, as it lacks variables, loops, and recursion, but it still allows for elaborate processing of SIP signaling.

3.4.8 CPL Example

To get some flavor of what CPL enables, consider the following example, inspired by some of the examples in RFC3880:

```
<?xml version="1.0" encoding="UTF-8"?>
<cpl>
 <subaction id="voicemail">
 <location url="sip:han@voicemail.ngnresearch.
com">
   <proxy />
  </location>
 </subaction>
 <incoming>
  </address-switch field="origin" subfield="user">
   <address is="anonymous">
     <reject status="reject"
         reason="No anonymous calls please"/>
   </address>
   <otherwise>
    <location url="sip:han@ngnresearch.com">
     <proxy timeout="10">
      <busy>
       <sub ref="voicemail" />
      </busy>
      <noanswer>
       <address-switch field="origin"
            subfield="host">
        <address subdomain-of="whitehouse.gov">
         <lookup source="http://www.
           ngnresearch.com/locate.php?user=han"
           timeout="8">
          <success>
           <proxy/>
          </success>
          <failure>
           <location url="tel:+34696858585">
```

```
            <proxy />
          </location>
        </failure>
       </lookup>
      </address>
      <otherwise>
        <sub ref="voicemail" />
      </otherwise>
     </address-switch>
    </noanswer>
    <failure>
     <mail url= mailto:han@ngnresearch.com?
           subject=You%20are%unreachable! />
    </failure>
   </proxy>
  </location>
  </otherwise>
 </address-switch>
 </incoming>
</cpl>
```

This example shows a CPL script that first defines a subaction consisting of proxying (i.e., forwarding) a SIP request to the destination `sip:han@voicemail.ngnresearch.com`. The script then proceeds to instruct the server what to do on an incoming session request.

First, the script tells the server to look at the user field of the origin of the SIP request (as specified by the `From:` header line of the SIP message). If the user is anonymous, then the server has to reject the session (by sending back a `603 Decline` SIP response on the reverse path to the originator).

If the call is not from an anonymous user, the server forwards the SIP request to `sip:han@ngnresearch.com` and sets a 10-second timeout for receiving the response.

If the UA at `sip:han@ngnresearch.com` answers that it's busy (by returning a `486 Busy here` response), then the script runs the voicemail subaction defined earlier that forwards the request to a voicemail server at `han@voicemail.ngnresearch.com`.

If the UA at `sip:han@ngnresearch.com` does not answer within the set time limit of 10 seconds, the script tells the server to check whether the SIP request came from `whitehouse.gov` (White House on the line!). If so, then the script tries to find user Han by looking up his current SIP address on the PHP application at URL `http://www.ngnresearch.com/locate.php?user=han`, which could be an application that keeps track of the SIP addresses of ngnresearch.com employees as they register in different domains (home, work, and so on) and on different devices (desktop PC, SIP phone, tablet, smart phone, and so on). If this lookup produces a SIP address, the script sends the SIP session there. If not, then the script makes a last ditch attempt to proxy the call to Han's mobile phone number `+34696858585` (a hard phone number, not a SIP address!). Calls that don't come from the White House are proxied to voice mail, by running the voicemail subaction defined earlier.

Finally, if the server receives any failure response from `sip:han@ngnresearch.com`, then the script sends an e-mail to `han@ngnresearch.com` with the subject "You are unreachable!".

3.5 MMTEL

In the previous sections we've seen that IMS provides a network for multimedia communications, using the versatile SIP protocol for signaling. IMS offers more flexibility than the public switched telephone network (PSTN) when it comes to providing multimedia communications and value-added services, but on the other hand the PSTN provides a tighter definition of service. One could say that the PSTN defines a network *and* its service (i.e., telephony, using standardized codecs and signaling procedures), while IMS defines only the enabling network.

To make services work across network operator boundaries, they have to interoperate. IMS allows operators to implement their multimedia communications in different ways using different SIP dialogs, different media flows, and different codecs, which makes interworking more of a challenge. This is where the PSTN has an edge and where the flexibility of IMS can actually be a handicap. To address this problem, operators around the world have sought to standardize basic IMS services so that they work across network boundaries.

The 3GPP and ETSI have defined IMS multimedia telephony (MMTEL) to ensure that subscribers can use basic IMS services such as voice and video calls, emergency calls, messaging, and fax across operator boundaries, and they can make voice calls between IMS and the PSTN. MMTel does not introduce any new IMS functionality or protocols. It merely prescribes the use of specific SIP flows, transport protocols, and codecs so as to ensure interoperability.

The 3GPP technical specifications 22.173 [14] and 24.173 [15] describe the MMTEL service and its implementation, respectively. MMTel offers services that include voice calling, video calling, conferencing, instant messaging, fax, file transfer, and sharing (images, video, audio clips). MMTEL defines quality of service aspects so as to ensure a uniform experience across network, as the PSTN does. The 3GPP also elaborately specifies MMTEL charging procedures in TS 32.275 [16].

Like many voice over IP services in the Internet, MMTel uses a presence service as the launch pad for multimedia sessions, which visualizes the status of subscribers and how to reach them. MMTel manages a subscriber's multimedia communications from a single SIP session. This allows a subscriber to add and drop multimedia flows (voice, video, messaging, file sharing) within a session and facilitates synchronization between different services (e.g., synchronization between audio and video in a videoconference).

Figure 3.20 shows the usual deployment of MMTel in IMS. The S-CSCF forwards MMTEL communications to a dedicated AS that handles MMTEL SIP signaling. An additional AS takes care of supplementary services such as conferencing, originating identification presentation, anonymous communication rejection, communication waiting (CW), communication barring (CB), completion of communications to busy subscriber (CCBS), completion of communications on no reply (CCNR), advice on charge (AOC), reverse charging, customized ringing signal (CRS), and others. Note that these services bear strong resemblance to services enabled by IN in the PSTN. In fact, one could liken the MMTel AS to the SSF and the supplementary AS to the SCF in CAMEL and IN (see Section 2.5.1).

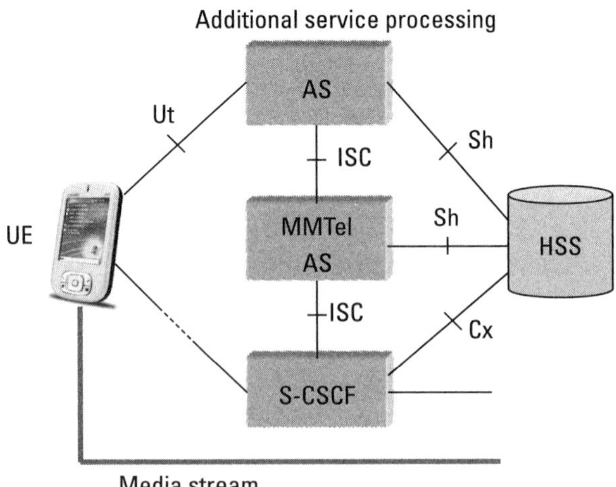

Figure 3.20   Deployment of MMTEL in IMS.

Figure 3.20 also shows that the AS has an XML configuration access protocol (XCAP) interface labeled Ut to the UE to let the subscriber manage these supplementary services.

As we've seen in Section 3.2.3, LTE has no circuit switched core network and uses MMTEL to offer voice telephony over IP.

3.6 Rich Communications Services

The rich communications services (RCS) published by the GSM Association (GSMA) have a similar intent as MMTEL, but they have a different background and use different (though similar) technology.

The first thing to observe is that the GSMA is an organization with a different scope, objectives, and structure than the 3GPP. GSMA is an industrial organization that represents the interests of mobile operators. It does not publish standards and should not be considered a standards development organization (SDO). RCS therefore does not introduce any new standards but merely endorses and profiles[13] existing standards from other organizations, principally the IETF and the Open Mobile Alliance (OMA). On the other hand, GSMA does maintain an interoperability testing (IOT) and certification program for RCS, managed by the Global Certification Forum (GCF). Certified vendors and service provides can use GSMA's Joyn brand for RCS, which testifies interoperability. GSMA also emphasizes the IP interconnect (IPx) infrastructure needed to connect IMS networks and ensure interoperability of RCS services between them.

At the time of writing of this book, the most recent version of RCS is version 5.1, which offers IP voice call, best effort video call, instant messaging, presence information, group chat, content sharing, file transfer, geolocation exchange, and capabilities exchange between RCS clients. RCS foresees embedded clients to be preinstalled on feature phones as well as downloadable clients that can be installed as an app on smart phones.

RCS uses standard IMS functions and interfaces, and does not prescribe any specific configuration of these. Figure 3.21 shows a typical deployment of RCS, including the IPx gateway between the networks that GSMA recommends.

---

[13] To profile a specification means to endorse or exclude certain options that the specification may offer or to select particular values for parameters in the specification. This narrows the specification down to more specific use in a particular context. An example of profiling is to specify that a protocol message must carry a certain header line, or that a session must use a particular codec.

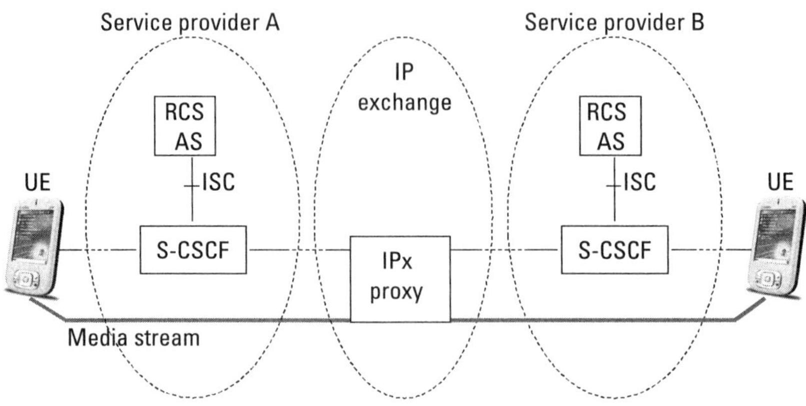

Figure 3.21　Typical deployment of RCS.

3.6.1　SIP for Instant Messaging and Presence

Technically, RCS hinges on the SIP for instant messaging and presence leveraging extensions (SIMPLE) specifications of the OMA. These specifications describe how to use SIP for exchanging instant messages and presence information between UAs.

SIMPLE specifies how SIP negotiates an instant messaging (IM) session between two participants and then uses the message session relay protocol (MSRP) to stream instant messages between them. The mechanism used here does not differ significantly from the way SIP negotiates any other session, except that the media stream is MSRP defined in IETF RFC 4975 [17]. Alternatively, SIP can also carry instant messages itself, taking advantage of the MESSAGE request described by IETF RFC 3428 [8].

SIMPLE also describes how a UA can subscribe to presence events, receive presence notifications, and publish presence updates using the SIP SUBSCRIBE, NOTIFY, and PUBLISH requests. Figure 3.22 shows the basics of this mechanism. The UA that publishes presence updates for a subscriber is referred to as the *presentity*,[14] while UAs that follow presence updates for a certain subscribers are called *watchers*.

The presentity uses the SIP PUBLISH request to send presence updates (such as "available," "not available," and "no calls, instant messages only") to the presence server in the RCS AS. A UA becomes a watcher by sending

---

[14] You won't find *presentity* in the English dictionary, but it is used by IETF and OMA to refer to the functional entity that publishes a user's presence information. One could read it as the contraction of *presence* and *entity*.

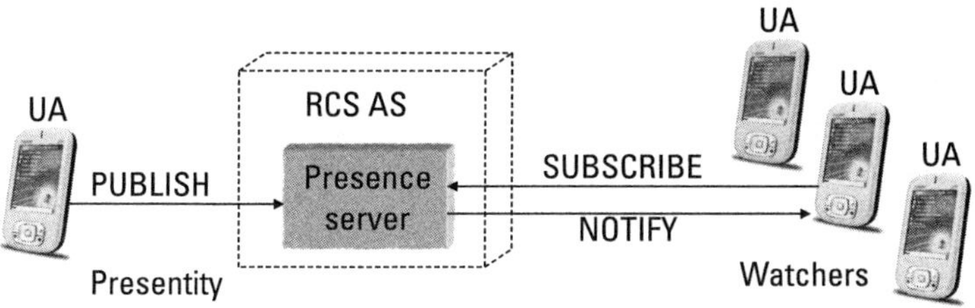

Figure 3.22 RCS use of SIMPLE to exchange presence information.

a `SUBSCRIBE` request for presence updates. When the presentity publishes a presence update, the presence server updates the watchers with a `NOTIFY` message.

3.6.2 Presence Information Data Format

SIMPLE uses the presence information data format (PIDF) defined in RFC 3863 [18] in the payload of the `PUBLISH` message to update the presence information for a subscriber. PIDF has a simple XML structure that allows for extensions with application-specific status information. The following is an example of a basic PIDF document:

```
<?xml version="1.0" encoding="UTF-8"?>
<presence xmlns="urn:ietf:params:xml:ns:pidf">
 <tuple id="fr38x3">
  <status>
   <basic>open</basic>
  </status>
  <contact> mailto:johan.zuidweg@ngnresearch.com</contact>
  <note xml:lang="en">I'm in a meeting</note>
  <note xml:lang="es">Estoy en reunión</note>
  <timestamp>2013-12-11T15:31:20Z</timestamp>
 </tuple>
</presence>
```

A PIDF document has a root element called `<presence>`, and within this element one or more `<tuple>` elements provides a piece of presence information. The `<status>` element in a tuple provides basic information on whether the subscriber's UA is open or closed for communication.

A tuple may also contain a `<timestamp>` to keep track of the timing of the presence update, a `<note>` qualifying the status, intended for the watchers, and `<contact>` information for an alternative means to contact the subscriber.

As SIMPLE and RCS provide several SIP-based communication services, SIMPLE has extended PIDF with a `<service-description>` element to qualify what service the presence information applies to. This allows a subscriber to set different presence information for different services. For example, the subscriber may state that he or she is available for instant messaging but unavailable for a voice call.

An example of such a service description identifying the instant messaging service, using the OMA-defined service identifier `org.openmobilealliance:IM-session`, is:

```
<service-description>
  <service-id>
    org.openmobilealliance:IM-session
  </service-id>
 <version>1.0</version>
<service-description>
```

SIMPLE defines further PIDF extensions to convey a subscriber's mood, willingness to receive session invitations, geolocation, device type, and other information.

Apart from SIMPLE, RCS uses other OMA specifications, such as XML document management (XDM) and converged IP messaging (CPM). RCS uses XDM to store and manage XML data, such as PIDF documents, and CPM to define the structure of the message store. We won't consider these in further detail here.

3.6.3 RCS-e

GSMA has specified a low entry version of RCS called RCS-e, which provides a more basic set of services intended for use with the legacy (circuit switched) voice telephony and short messaging services. The four services supported by RCS-e are instant messaging and chat, file transfer, image sharing, and video sharing. RCS-e does not require a presence server and uses the SIP OPTIONS request to discover the capabilities of subscribers.

To lower the bar for deploying RCS even further, some vendors and service providers will host RCS services for exploitation by others. They will provide IMS and RCS as cloud services ready for use by operators and service providers who are not ready to install an IMS or RCS system themselves.

Although operators worldwide and the GSMA are making a strong case for RCS, the industry has also criticized the RCS concept. The RCS specification process has been cumbersome because of the diverging interests of operators in different regions. The split between RCS and RCS-e is a symptom of this discord.

Some analysts argue that RCS stands no chance against well-established OTT services that offer voice and video over IP, instant messaging, group chat, and content sharing free of charge. RCS proponents point out that RCS works also on feature phones that cannot run voice over IP applications such as Skype or Google Talk, but critics do not believe that this represents a compelling business case. The future of RCS, if not of the entire telecommunications industry as we know it, remains uncertain.

3.7 IMS Implementations

Many vendors of telecommunications equipment offer IMS products, including smaller niche players. The Berlin-based Fraunhofer institute for Open Communications Systems (FOKUS) has created an IMS reference implementation that is available as open source, called the IMS Open Core. Although the Open IMS Core implements only the S-CSCF, P-CSCF, I-CSCF, and HSS functions of the IMS, it is sufficient for trying out basic call setup routines in IMS. As it is open source, the developer can easily connect it to application servers and other IMS elements to create more elaborate configurations.

You can download the Open IMS Core and its documentation from `http://www.openimscore.org`. This website also allows developers to become involved in the project and contribute comments, ideas, and code.

References

[1]   Tanenbaum, A. J., and D. Wetherall, *Computer Networks*, 5th Edition, Boston: Pearson Education, 2011.

[2]   Stallings, W., *Data and Computer Communications*, 10th Edition, Upper Saddle River, NJ: Prentice-Hall, 2013.

[3]   Wilson, P., *Internet Addressing in the 2010s, IPv4 Exhaustion and Address Transfers, and Their Impact on IPv6 Deployment*, Asia Pacific Network Information Center (APNIC), 2013.

[4]   Rosenberg, J., G.Camarillo, A. Johnston, J. Peterson, H. Schulzrinne, et al., "SIP: Session Initiation Protocol," RFC 3261, IETF, June 2002.

[5]   Handley, M., V. Jacobson, and C. Perkins, "SDP: Session Description Protocol," RFC 4566, IETF, 2006.

[6]   Roach, A. B., "Session Initiation Protocol (SIP)-Specific Event Notification," RFC 3265, IETF, 2002.

[7]   Niemi, A., "Session Initiation Protocol (SIP) Extension for Event State Publication," RFC 3903, IETF, 2004.

[8]   Campbell, B., J. Rosenberg, H. Schulzrinne, C. Huitema, and D. Gurle, "Session Initiation Protocol (SIP) Extension for Instant Messaging," RFC 3428, IETF, 2002.

[9]   Noldus, R., U. Olsson, C. Mulligan, I. Fikouras, and A. Ryde, et al., *The IMS Application Developers's Handbook*, Oxford, UK: Academic Press, 2011.

[10]   Nakamura, M., and S. Reiff-Marganiec (eds.), *Feature Interactions in Software and Communication Systems X*, Leiden, Netherlands: IOS Press, 2009.

[11]   Third Generation Partnership Project; Technical Specification Group Services and System Aspects; IP Multimedia Subsystem (IMS); Stage 2, 3GPP TS 23.228.

[12]   Kocan, K. F., W. D. Roome, and V.Anupam, "Service Capability Interaction Management in IMS Using the Lucent Service Broker™ Product," *Bell Labs Technical Journal*, Special Issue: The IP Multimedia Subsystem, Vol. 10, No. 4, 2006, pp. 217–232.

[13]   Lennox, J., X. Wu, and H. Schulzrinne, "Call Processing Language (CPL): A Language for User Control of Internet Telephony Services," RFC 3880, IETF, 2004

[14]   Third Generation Partnership Project; Technical Specification Group Services and System Aspects; IP Multimedia Core Network Subsystem (IMS) Multimedia Telephony Service and Supplementary Services; Stage 1, 3GPP TS 22.173.

[15]   Third Generation Partnership Project; Technical Specification Group Core Network and Terminals; IMS multimedia telephony communication service and supplementary services; Stage 3, 3GPP TS 24.173.

[16]   Third Generation Partnership Project; Technical Specification Group Services and System Aspects; Telecommunication management; Charging management; MultiMedia Telephony (MMTel) charging, 3GPP TS 32.275.

[17]   Campbell, B., R. Mahy, and C. Jennings, "The Message Session Relay Protocol (MSRP)," RFC 4975, IETF, 2007

[18]   Sugano, H., S. Fujimoto, G. Klyne, A. Bateman, and W. Carr, et al., "Presence Information Data Format (PIDF)," RFC 3863, IETF, 2004

[19]   Fajardo, V., J. Arkko, J. Loughney, and G. Zorn, "Diameter Base Protocol," RFC 6733, IETF, 2012.

4

Programming Communications Applications and Services with Java

The previous chapters have discussed circuit and packet switched, fixed, and mobile network architectures for voice and multimedia communications. IN defines how to compose services from service independent building blocks (SIBs), but does not explain how to program them or how to integrate IN services with back office systems. IMS specifies the role of the AS in enabling value-added services but does not say what goes on inside an AS.

In this section we'll study Java, one of the most prolific programming languages of our time, and see how programmers use it to create applications and services for telecommunications and for IMS in particular. Since you may not be a programming expert, we'll start this chapter with a short tutorial of object-oriented programming and Java. We won't go into too much technical detail since there are excellent books on Java programming, but we'll see just enough to understand how object-oriented programming with Java works.

4.1 Object-Oriented Programming

Imagine having to program an air traffic control system, an accounting system for a large bank, or a full-blown IMS system. These are immensely complex tasks that require a structured and disciplined approach.

Programming is the art of breaking a complex problem down into smaller pieces, and these pieces into even smaller pieces, until you have bite-size problems to which you can apply straightforward solutions such as reading input from a keyboard, displaying something on a screen, retrieving something from a database, or sorting a list.

There are many ways to program complex tasks. A classical approach is procedural programming, which builds the solution to a problem as a series of smaller programs called procedures or subroutines executed in sequence by the main program. Fortran, Pascal, and C are examples of procedural programming languages. Functional programming (e.g., LISP) and logic programming (e.g., PROLOG) provide other approaches to solving a complex problem.

The most widely used type of programming today is object-oriented programming, which defines the problem space and its solutions in terms of objects. An object represents a set of data and operations on that data. Since the late 1980s, object-oriented programming has become enormously popular for various reasons:

- Objects provide an intuitive way of representing a problem and its solutions.
- Object-oriented programming promotes modular software development by dividing the program up into groups of objects that can be programmed independently. This makes it easier to distribute the work over teams and facilitates testing and integration.
- Object-oriented programming makes it easy to reuse code for well-known problems. Programs generally tend to consist of standard solutions to well-known problems such as sorting a list, storing a value in a database, or generating a pie chart from a series of numbers. It makes no sense reprogramming such routine tasks if you can reuse a piece of code already written that does the trick.

In the following sections, we'll take a closer look at objects and how to design software with them.

4.1.1 Classes and Objects

Since the late 1980s, authors have published hundreds of books and papers on programming with classes and objects ranging from academic to practical. Although all programmers know what a class and an object is, different publications sometimes give quite diverse definitions of these terms. The concept of classes and objects is more elusive than may seem at first sight, so we'll start

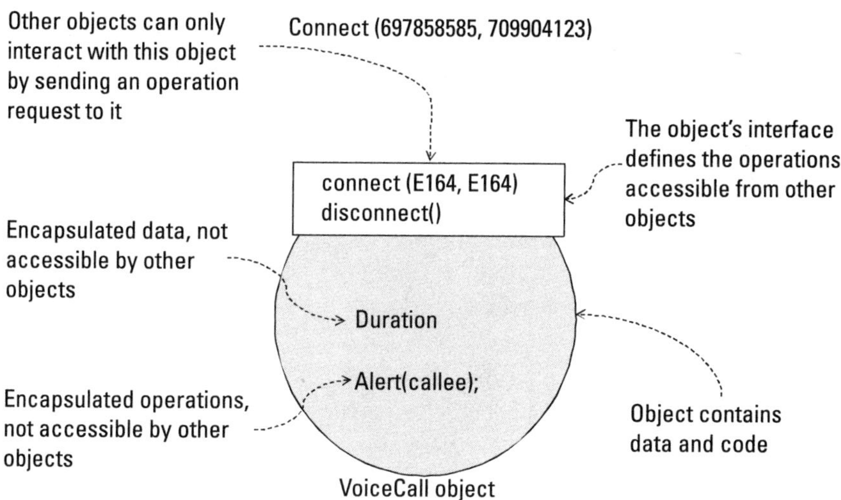

Figure 4.1   Object concept.

this chapter by getting a few definitions and concepts straight before talking about the object-oriented programming language Java.

An object consists of two things: data and the operations[1] you can perform on that data. An object encapsulates its data, meaning that only an object's operations can manipulate the object's data. A program can only access or manipulate an object's data through the object's operations.

A program interacts with an object by sending a message to it, requesting the execution of one of the object's operations. When an object receives such a request, it executes the operation with the parameters specified in the message and returns a message with the result. Figure 4.1 illustrates the concept of an object, how it encapsulates data, and how it responds to operation requests.

Objects exist as run time artifacts. They manage memory space for their data and keep the code for their operations. To program an object, an object-oriented programming language offers two constructs that tell the computer how to implement objects at run time:

- An *interface* defines the operations supported by an object.
- A *class* describes the implementation of an object's operations and internal data.

---

[1] The terms *method* or *function* are also used commonly to refer to operations on an object. In this book we'll mostly use the more formal term *operation*.

Classes and interfaces have a close relationship because a class has to implement the operations defined by an object's interface. Here is an example of how to specify an interface for the `VoiceCall` object in Figure 4.1 in Java[2] (assuming that the `E164Num` type has been declared previously to represent E.164 numbers):

```
interface VoiceCall {
  void connect(E164Num caller, E164Num callee);
  void disconnect();
}
```

And here is an example of a class called `VoiceCallImpl` that implements the `VoiceCall` interface:[3]

```
public class VoiceCallImpl implements VoiceCall {
  private Duration d;   // this variable is local
                        // to the class and cannot
                        // be accessed from outside
                        //it
  public void connect(E164Num caller, E164Num
callee) {
    /*here goes the Java code to connect caller
and callee*/
  };
  public void disconnect() {
    /*here goes the Java code to disconnect the
parties*/
  };
  private alert (E164Number destination){
    /*this operation is local to the class and
is not part of the interface*/
  };
```

---

[2] If you don't know Java or haven't programmed before, you may have some trouble deciphering this and understanding what void means. Don't worry too much about it at this point.

[3] This example does not show particularly brilliant programming. In fact you can challenge the way I've chosen to program a voice call with objects here. For example, what does it mean for a call object to exist before it has connected two parties or after the parties have disconnected? Don't look for too much behind these bits of incomplete code; their only intention is to provide the simplest possible examples of some of the basic principles of object-oriented programming.

4.1.2 Inheritance

Object-oriented programming allows you to create interfaces and classes by building on existing interfaces and classes. In other words, an interface can inherit the operations of an existing interface and modify them or extend them with new operations. Likewise, a class can inherit the code of an existing class and build on this. Inheritance is a powerful feature that enables the reuse of functionality and code.

Note that the previous paragraph mentions interface inheritance and class inheritance separately because they serve different purposes:

- *Interface inheritance* lets you substitute a class by another, as the subclass provides the same interface as the superclass. Interface inheritance is about the exterior of an object (its interface).
- *Class inheritance* lets you reuse the program for an existing class and extend it to create a new class with additional or modified functionality. Class inheritance is about the interior of an object (its data and code).

Building on the example from the previous section, the following Java example tells the computer that objects of type `MultiPartyCall` have all the functionality of a `VoiceCall` object, plus new functionality to add and drop parties in the call:

```
interface MultiPartyCall extends VoiceCall {
   void addParty(E164Num p);
   void dropParty(E164Num p);
}
```

A program can use a `MultiPartyCall` object anywhere where a `VoiceCall` is needed, because a `MultiPartyCall` is also a `VoiceCall`.

The following Java example tells the computer that `MultiMediaCallImpl` objects reuse the code of `VoiceCallImpl` objects and adds two additional operations, `addMediaStream` and `removeMediaStream`:

```
public class MultiMediaCallImpl extends
VoiceCallImpl {
   public void addMediaStream(Media m) {
      // here goes Java code to add a media stream
   };
   public void removeMediaStream(Media m) {;
```

```
    // here goes Java code to remove a media stream
  };
}
```

Interface inheritance tells the computer to reuse *what* an object provides while class inheritance tells the computer to reuse *how* the object does things.

4.1.3 Composition

Inheritance gives the programmer a static way of reusing definitions and code at the time of writing the program. A program can also compose objects from other objects at runtime. As an object can contain any kind of data, including other objects, a program can dynamically build complex objects from simpler ones at run time. For example, our `VoiceCallImp` class from Section 4.1.1 can contain a variable of another class to represent a connection between two parties:

```
public class VoiceCallImpl implements VoiceCall {
  // create a new connection object
  Connection c = new Connection();
  /*(here goes the rest of the code for
VoiceCallImpl
  /*which can do something with connection c*/
}
public class Connection {
  /*the code that describes objects of class
Connection*/
}
```

This piece of code tells the computer that a `VoiceCallImp` object can contain a `Connection` object and do things with it. This way, the program can compose objects from others and have objects manipulate other objects at runtime.

Object-oriented programming uses intuitive and powerful concepts, but this does not make object-oriented programming easy. The previous examples have probably told you that already. When writing an object-oriented program, one has to make difficult decisions and tradeoffs such as:

- How to represent your application and data by objects (encapsulation);
- How big or how small to make the objects (granularity);
- To what degree to reuse existing code (reusability and flexibility).

4.1.4 Design Patterns

Programmers joke that the most widely used programming tool is copy-paste. In fact the reuse of patterns that have proven to work is common in many areas, including fiction ("Cinderella meets prince," "detective unites suspects and resolves murder case"), and cinema ("superhero saves the world," "good fights bad and wins"). In the 1970s, the architect Christopher Alexander proposed the reuse of proven and practical designs in architecture and created a pattern language to describe solutions to well-known architectural problems. Alexander's work prompted computer scientists to start applying the concept of design patterns to object-oriented programming in the 1980s. In 1994, four computer scientists published a book that documented the use of design patterns in object-oriented programming and contained a reference guide of patterns in use by that time. Their book, *Design Patterns: Elements of Reusable Object-Oriented Software* [1], became the best-selling book in computer science of all time. Design patterns are now commonly used in object-oriented design and programming.

In essence, a design pattern is a reusable solution to a well-known problem. The description of a design pattern assigns a name to the pattern for reference, defines the problem, and presents the reusable solution. In addition, a design pattern may identify tradeoffs and consequences of applying the solution, as well as alternatives and variations of the problem and its solution.

Design patterns come in many shapes and flavors depending on the problem they are solving and how they approach it. There are several ways to classify design patterns. One way is to distinguish between static and dynamic design patterns:

- *Class patterns* are static and define classes in terms of other classes by inheritance.
- *Object patterns* are dynamic and describe how objects interact with other objects at run time to solve a problem.

Another way to classify design patterns is by the type of problem they solve:

- *Creation patterns* describe how classes and objects create and initialize other objects.
- *Structural patterns* describe how to define classes (design time) and build objects (run time) from other classes and objects.
- *Behavior patterns* describe how to build a control flow with classes (design time) or with communicating objects (run time).

Table 4.1
A Classification of Design Patterns

	Creation	**Structure**	**Behavior**
Class (static)	Describe how classes instantiate objects	Define classes by inheritance	Describe a control flow by inheritance
Object (dynamic)	Describe how objects create other objects at run time	Compose objects from other objects at run time	Build a control flow from communicating objects at run time

Table 4.1 combines these classifications into a single view. Of course this is not the only way to classify design patterns; one can classify them by other criteria too.

Although designers and programmers have defined hundreds of design patterns, they share underlying principles that characterize good practice in object-oriented programming. Some of the key principles that lie at the basis of almost all design patterns are as follows:

- *Design for change.* Any programmer will tell you that change defines the art of programming. User requirements may change. Hardware, operating systems, and libraries may change. Your program may need future extensions. The more your design anticipates such changes, the less pain and cost these changes will cause, especially when the project is at an advanced stage. This principle underlies almost all the other principles that follow.
- *Encapsulate what varies.* Any implementation that is subject to change is best encapsulated behind a more stable interface, so as to reduce the dependencies on volatile code or data.
- *Program to interfaces, not implementations.* In Section 4.1.1 we've seen that interfaces define the external behavior of objects. Programming to interfaces allows the programmer to change the internal implementation of an object or to substitute an object with another object of a subclass without the need to reprogram the objects that depend on the interface. This promotes modular design and programming. For example, a programmer should resist the temptation to add any public operations to the `VoiceCallImp` class in Section 4.1.1 that are not in the `VoiceCall` interface and call them from other objects.

- *Delegate tasks whose implementation can change.* If a class has an operation that has variable implementation in subclasses, it is better to take that operation out of the class and delegate it to another class designed specifically for the purpose of implementing the volatile functionality. This way, if you need to change the functionality you only need to reimplement that specific class.
- *Favor composition over inheritance.* Inheritance is a powerful but static construct. In many cases, the dynamic composition of objects from other objects at run time provides more flexibility and better resilience to change (see the earlier point about designing for change).

4.1.5   Example: The Observer Pattern

The Java implementations of IMS and SIP based systems make frequent use of design patterns. This section will not attempt to describe all the design patterns you may need when programming IMS and SIP, as this would divert us too much from the main topic of the chapter and of the book. There are many excellent reference works on design patterns, one of them the now-20-year-old but timeless book by the so-called Gang of Four [2].

We'll describe one design pattern here that applies to the Java implementations of SIP and IMS that we'll see later in this chapter, the observer pattern. The observer pattern provides a design for unidirectional many-to-many communication between producers and consumers of information that does not require direct contact between producers and consumers. This way of communicating also goes by the name of *publish-subscribe* or *event-driven* communication. Figure 4.2 shows the observer pattern.

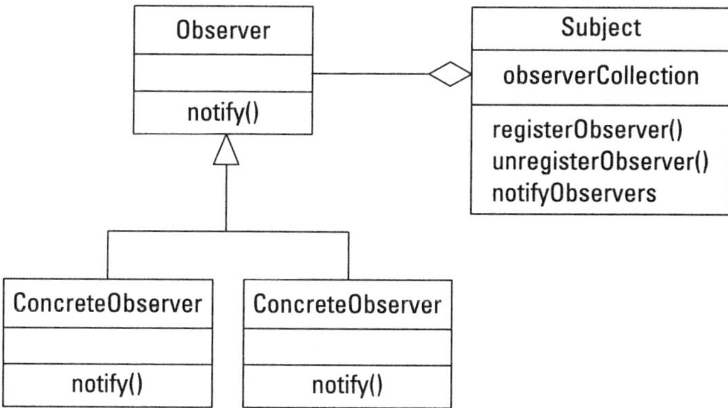

Figure 4.2   The observer pattern.

The observer pattern has three principal classes:

- `Subject` represents the asynchronous communication channel between observers and the producer of the information. Observers subscribe to receiving information from the `Subject` by calling the `registerObserver()` operation. `Subject` maintains a list of all registered subscribers in the attribute `observerCollection`. A producer publishes information by calling the `notifyObservers()` operation on the `Subject`, sending the information as parameter of the operation. The `Subject` pushes this information to all registered observers by calling the `notify()` operation (see the next bullet) on each registered observer.
- `Observer` defines the interface of the information consumer with one operation, `notify()`.
- `ConcreteObserver` classes let the programmer implement the specifics of the `notify()` operation; in other words, they let the programmer decide what each `ConcreteObserver` has to do with the information received.

Designers and programmers commonly use the Observer pattern to implement asynchronous communication. Some literature refers to the `Subject` class as the topic (in publish-subscribe systems) or event channel (in event-driven communication).

Now that we've reviewed some of the principles of object-oriented design and programming, we'll take a closer look at Java, one of the most widely used object-oriented programming languages.

4.2   Programming with Java

Section 4.1 has given us a preview at some bits of Java code, but what is Java really, and what makes it so popular? Before the creation of Java in the mid-1990s, C++ was one of the most commonly used object-oriented language for general-purpose programming. As the name suggests, C++ extends C, a procedural programming language traditionally deployed on Unix systems. While C++ provides a powerful object-oriented extension to C, it also has its handicaps. By definition, C++ is an extension of C, so each C program is also a C++ program. C++ allows a programmer to mix object-oriented programming with lower level, nonobject-oriented C programming. Critics argue that not enforcing object-oriented programming encourages the creation of bad software. Writing a well structured C++ program requires a lot of discipline, and C++ has a steep learning curve.

Sun Microsystems (acquired by Oracle in 2010) created Java in the mid-1990s to provide a language that preserves the best of C++ but mitigates some of its disadvantages. Despite the similarity in name, Java bears very little in common with JavaScript.[4]

Java imposes object-oriented programming more strongly than C++ and does not give a programmer low-level access to a computer's memory as C does. This improves program structure and reduces the chances of writing faulty code. Java also addresses the more general problem that programming languages need different compilers for every processor and operating system by introducing the notion of the virtual machine.

4.2.1 Java Virtual Machine

A virtual machine consists of software that emulates an imaginary processor and runs what is called *byte code*. The virtual machine and its byte code strongly resemble a physical processor and its instruction set. One can consider the virtual machine a common denominator for real-life processors, a software defined processor that behaves as physical processors do but without providing exactly the set of instructions of any existing processor.

With the Java virtual machine, the execution of a Java program requires two steps. First, the Java compiler translates the program to byte code for the virtual machine. The virtual machine then interprets the byte code on a physical processor; in other words, the virtual machine translates the byte code to machine instructions for the processor in real time, instruction by instruction.

This approach has the enormous advantage that Java needs only one compiler from Java to byte code and that the byte code consistently gives the same results on each virtual machine. A compiled Java program can run on any physical computer that has the Java virtual machine installed on it. Oracle calls this "write once, run anywhere™," and it has made programming across computer systems much easier.

---

[4] JavaScript is an interpreted language for execution in a browser. Although it is object-oriented and superficially resembles Java syntax, it does not compile to byte code, does not run on a virtual machine, and executes very differently from Java. The name *Java* in JavaScript is a mere coincidence that creates more confusion than anything else. JavaScript is of no immediate relevance in this chapter, so we'll leave it aside for the moment.

The price to pay is that the execution of byte code requires a virtual machine on the computer. Figure 4.3 compares the process of normal compilation with compilation into byte code for a virtual machine.

The execution of byte code on a virtual machine can still give slightly different results on different processors, but on the whole the execution of Java code becomes more uniform and predictable than when compiling directly to object code.

A virtual machine brings a certain penalty in performance, as it must translate each byte code instruction into a machine instruction before the processor can execute it. For applications that require high performance or that have to exploit specific features of a processor, it is also possible to compile Java directly to object code.

4.2.2   Java API

Java does not only provide a compiler and a virtual machine, it also has a library of preprogrammed classes the programmer can use to perform common

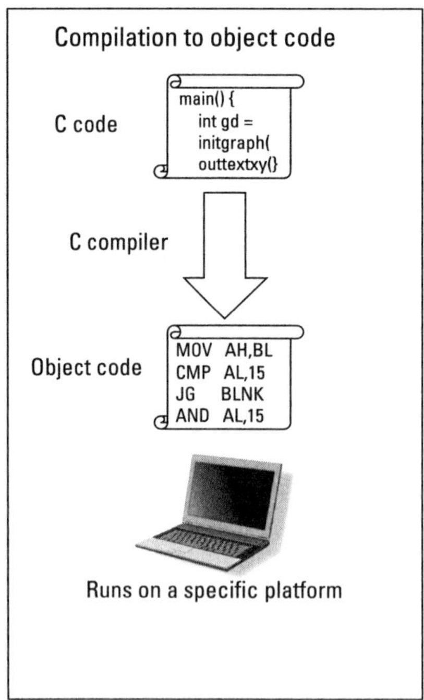

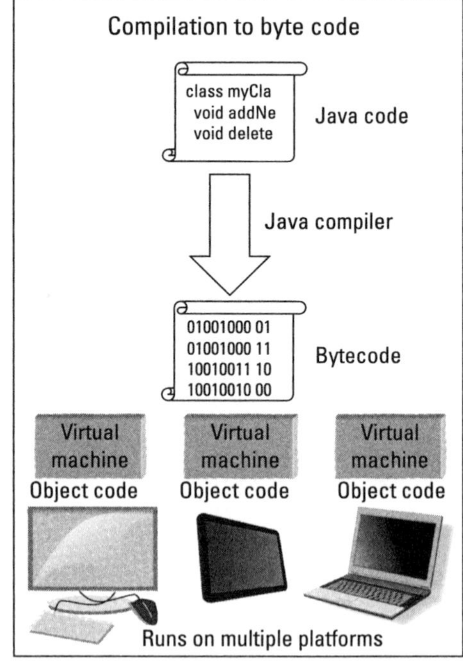

Figure 4.3   Compilation to byte code.

tasks such as comparing two strings, opening a TCP/IP socket, or displaying a window on the screen. This library of preprogrammed utilities is commonly referred to as the Java application programming interface (API).

The quality and comprehensiveness of the Java API has no doubt contributed to Java's popularity as a programming language. The Java API provides utilities for such widely varying tasks as manipulating images, handling sockets, connecting to structured query language (SQL) databases, and, as we'll see further on in this chapter, SIP processing.

The Java API comes in packages that provide sets of related classes for a specific purpose. For example, the `java.awt.print` package provides utilities for printing and `java.net` provides utilities for networking, including TCP/IP sockets. Whereas some packages such as `java.lang` apply to (almost) all programs, not all applications need all packages. The `import` declaration in Java lets the programmer specify the packages needed for a particular program.

A programmer needs APIs to be able to use communications protocols such as TCP/IP and SIP within a Java program. Figure 4.4 shows the relationship between an API and the protocol it implements.

The API provides the programmer with classes and operations to initialize and stop the protocol stack, send and receive protocol messages, and maintain the state of a protocol dialog. A particular protocol can have different APIs with varying degrees of granularity. Some APIs offer low-level access to a protocol, allowing the programmer to create, send, receive, and process individual protocol messages. Others offer a more abstract object-oriented view of the services provided by the protocol.

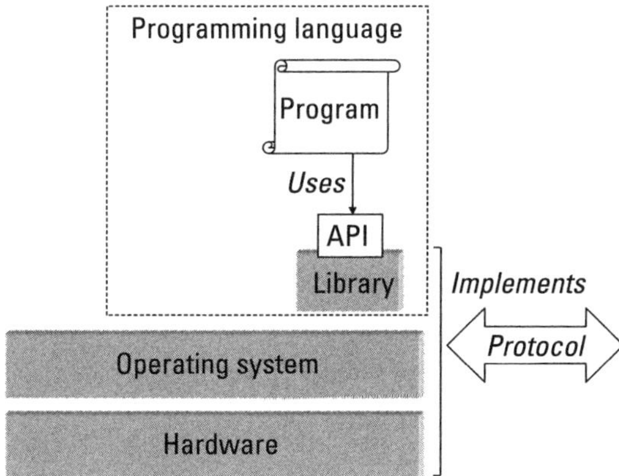

Figure 4.4 Relationship between API and protocol.

4.2.3 Java Platforms

Oracle provides three Java platforms for different programming environments. Each platform comes with a virtual machine and API adapted to the application environment:

- Java Standard Edition (JSE) provides the standard Java virtual machine and API needed for common programming tasks.
- Java Enterprise Edition (JEE) provides an extended API and richer run time environment that is a superset of Java SE. JEE provides additional facilities for building large-scale programs with strong reliability and security features, and the ability to connect to legacy databases.
- Java Micro Edition (JME) provides a lightweight run time environment and reduced API for applications that have to run in resource-limited devices such as mobile phones, media players, and wearable devices. JME provides a subset of the JSE API but also provides specific packages for lightweight applications.

JEE usually comes configured with the following two components:

- The *connected limited device configuration* (CLDC) [3] specifies the Java virtual machine functionality and basic libraries that devices with constrained resources must support.
- The *mobile information device profile* (MIDP) [4] provides packages of higher-level libraries that run on CLDC and allow programmers to manipulate user interfaces (`javax.microedition.lcdui`), data storage (`javax.microedition.rms`), data connections, short messages, multimedia messages (`javax.microedition.io`), certificates, and encryption keys (`javax.microedition.pki`). A Java program that runs on MIDP is called a MIDlet.

Instead of CLDC, JEE can also use the connected device configuration (CDC) run time environment, which contains the full feature Java virtual machine but runs only on devices with sufficient computing resources and peripheral capabilities.

JEE also defines additional profiles (i.e., packages) that run on CLDC (and CDC) such as the information module profile (IMP) for embedded communications modules without screen and keyboard and the set top box profile (STBP) for television set top boxes. Whereas mobile devices widely support MIDP, IMP and STBP have less widespread use.

4.2.4 Java SDK

To run applications programmed in Java, a computer needs to have the Java virtual machine and API installed. Those who actually program Java applications will also need the Java compiler to create byte code that can run on the virtual machine.

In theory, a Java programmer can use any text editor to write his or her program. But in practice it requires enormous discipline and effort to indent each line, nest each statement, and reference each variable correctly, just to mention a few things a programmer has to worry about when typing out his or her program. A Java editor provides these facilities to the programmer and makes his or her life a whole lot easier.

The Java software development kit (SDK) provides the tools that programmers need to create their programs, such as the Java compiler, a Java editor, and debugging tools that help the programmer find errors in the code. Of course, the Java SDK also includes the virtual machine and API so the programmer can try out and debug his or her code.

4.2.5 Java APIs for Communications Applications

Now that we've seen the basics of Java, it's time to dive into how to use the language for creating communications applications and services. The Java API provides several packages for working with SIP and IMS with different degrees of abstraction. The remainder of this chapter will study them in more detail.

- Section 4.3 introduces the Java SIP (JSIP) API, a low-level API for sending, receiving, and manipulating SIP messages.
- Section 4.4 presents SIP servlets, which allow applications to create and respond to SIP messages in a similar way as HTTP servers process HTTP requests.
- Section 4.5 describes a run time environment and API for building carrier grade communications services, called Java API for Integrated Networks (JAIN) Service Logic Execution Environment (SLEE).
- Section 4.6 introduces the Java IMS API, an API for JEE intended for creating IMS applications on terminal devices.

4.3 JSIP API

The JSIP API [5] provides classes and operations for creating, parsing, and sending and receiving SIP messages. It provides the programmer with a low-level

API for manipulating SIP messages. The JSIP API assumes the installation of a SIP stack. A SIP stack is software that

- Parses incoming protocol messages and assembles outgoing protocol messages;
- Manages the transport of the SIP protocol messages over the physical network using TCP/IP or UDP/IP;
- Keeps the state of SIP transactions, relating responses to requests;
- Keeps the state of SIP dialogs (e.g., the INVITE dialog we've seen in Figure 3.8).

The JSIP API can work with open or proprietary SIP stacks programmed in Java or in other programming languages. If the JSIP API uses a SIP stack programmed in another language, it will need to provide Java wrappers[5] for the non-Java SIP stack functions.

4.3.1   JSIP API Structure

Figure 4.5 illustrates the main communication structure of the JSIP API, which uses a variation of the Observer design pattern we've seen in Figure 4.2.

The `SipProvider` interface of the JSIP API provides the main interface to the SIP stack and represents the subject of the Observer pattern. Applications have to implement the `SipListener` interface, which plays the role of the observer of incoming SIP messages. The SIP provider notifies the application of an incoming SIP request by calling the `processRequest()` operation on its `SipListener` interface. Likewise the `processResponse()` operation notifies the application of an incoming SIP response.

The application sends outgoing SIP messages by directly calling the `sendRequest()` or `sendResponse()` operation on the `SipProvider` interface. These operations carry a `Request` or `Response` object as parameter that specifies the request or reply to be sent by the SIP stack.

The application can also manage SIP transactions (request-response pairs) and dialogs (such as the INVITE three way handshake with intermediate responses; see Section 3.3.6) through the `SipProvider` interface.

4.3.2   Request and Response Handling

The application handles incoming SIP requests asynchronously. The `processRequest()` and `processResponse()` operations on the `SipListener` interface notify the application of the arrival of a SIP message but do

---

[5] A wrapper is a class that implements no functionality of its own but calls operations provided by other classes or interfaces to other, non-object-oriented software.

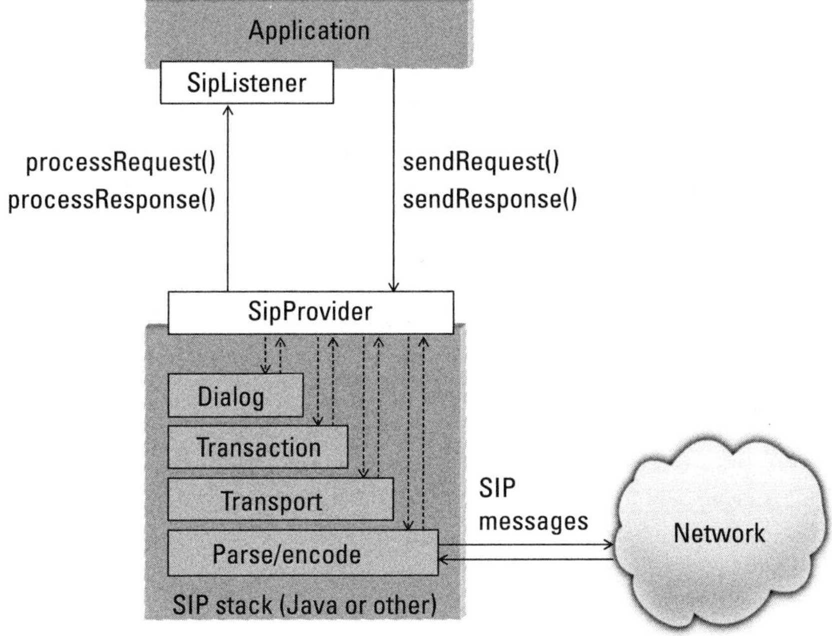

Figure 4.5 JSIP API main structure.

not contain the request or reply itself. It is up to the application to fetch the `Request` or `Reply` object describing the incoming SIP request or reply from the SIP provider and to process it. The JSIP API provides the `RequestEvent` interface for this purpose. Figure 4.6 illustrates how the notification of the incoming SIP message and its processing are decoupled.

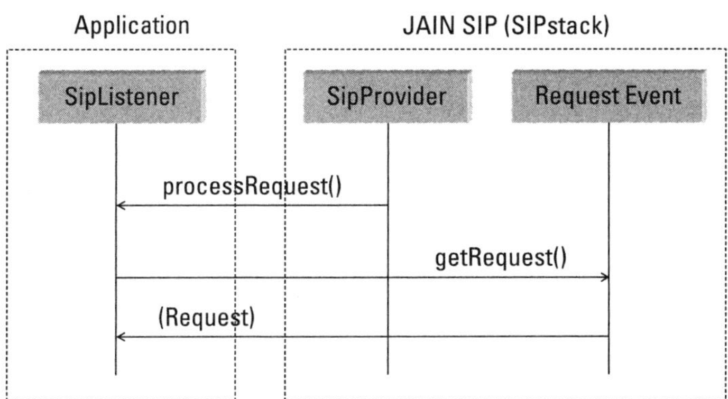

Figure 4.6 Asynchronous processing of incoming SIP messages in the JSIP API.

The SIP provider notifies the application of an incoming request by calling the `processRequest()` operation on its `SipListener` interface. The application then fetches the `Request` object by calling the `getRequest()` operation on the `RequestEvent` interface of the JSIP API.

The JSIP API builds `Request` and `Response` objects by composing them from `Header` objects to specify headers and `Address` objects to specify addresses. The Header class has subtypes for specific types of header lines, such as `To:`, `From:`, and `Via:` header lines (see Section 3.3.4). The `Address` class can contain a SIP identifier, a telephone URL, or an URI of any other type.

4.3.3   JSIP Interfaces

The previous section has introduced the two main JSIP interfaces, `SipProvider`, which encapsulates the SIP stack, and `SipListener`, which allows an application to receive SIP events. The JSIP API provides additional interfaces and classes for initializing the SIP stack, creating SIP requests and responses, and specifying addresses. The JSIP API provides the following interfaces to initialize the SIP stack and its `SipProvider` interface:

- `SipStack` is the interface for starting and stopping the SIP stack software, creating a `SipProvider` object and creating `ListeningPoint` objects.
- `ListeningPoint` represents a TCP/IP or UDP/IP socket to be used by the SIP stack to send and receive SIP messages. The `ListeningPoint` allows the application to get and set the transport parameters for the SIP stack, such as the IP address, type of transport (TCP or UDP), and TCP/UDP port. A SIP stack can use more than one `ListeningPoint` at a time.

In addition, the JSIP API provides factory interfaces to bootstrap itself and to create the objects mentioned previously:

- `SipFactory` is the initial factory that creates the SIP stack and the other factories. The JSIP API defines the `SipFactory` as a singleton (i.e., a static[6] class that creates exactly one `SipFactory` object).

---

[6] A static class allows the program to use operations of the class without instantiating (i.e., creating an object of) the class first. Among others, this facilitates bootstrapping, because it allows a class to create an instantiation of itself. It also enables the Singleton design pattern, which ensures that a class has no more than a single instantiation at all time.

- MessageFactory creates SIP requests and responses.
- HeaderFactory creates SIP headers.
- AddressFactory creates addresses such as IP addresses and SIP identifiers.

By now you may have lost track of all the classes the JSIP API uses and how they relate. Figure 4.7 shows a simple map of which objects create which. The application uses the SipFactory class to bootstrap and create the SipStack. Once it has a reference to the SipStack, the application can request the SipStack to create a ListeningPoint and a SipProvider. The application can now start sending and receiving SIP messages and build requests, responses, headers, and addresses through their respective factory interfaces.

Initialization of the JSIP API looks something like this in Java (slightly simplified):

```
/*Bootstrap SipFactory using the Singleton
pattern:*/
sipFactoryInstance = SipFactory.getInstance()
/*Create SipStack using name lookup (details
omitted) */
sipStackInstance=sipFactoryInstance.
createSipStack(name) */
```

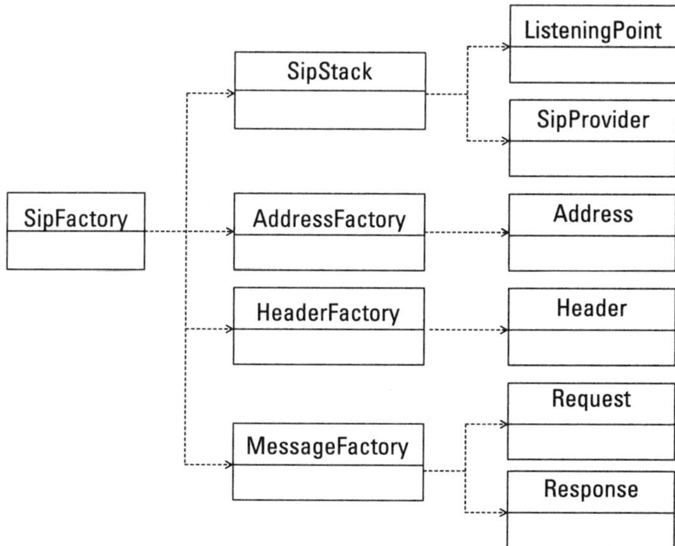

Figure 4.7 Main JSIP API objects and their creation.

```
/*Create ListeningPoint on which the SipStack has
to communicate*/
listeningPointInstance=sipStackInstance.
createListeningPoint(
                        ipAddr, udpPort,"udp")
/*Create SipProvider on the ListeningPoint
specified above*/
sipProviderInstance=sipStackInstance.
createSipProvider(
                    listeningPointInstance)
/* Register this application as SpiListener (i.e.,
observer) to the SipProvider */
sipProviderInstance.addSipListener(this);
/*Create the message, header and address
factories*/
messageFactoryInstance=sipFactoryInstance.
createMessageFactory;
headerFactoryInstance=sipFactoryInstance.
createHeaderFactory;
addressFactoryInstance=sipFactoryInstance.
createAddressFactory
```

Note that `SipFactory` uses the Singleton pattern to bootstrap: its static `getInstance()` operation returns the `SIPFactory` instance if it exists; otherwise, it creates one and returns the instance. The rest of the lines are more self explanatory.

4.3.4　Sending SIP Messages

After the initializations described in Section 4.3.3, the application can start sending and receiving SIP requests and responses. Figure 4.8 shows how the application can build and send a SIP request through the `MessageFactory` and `SipProvider` interfaces.

The application first creates a `Request` object by calling the `createRequest()` operation on the `MessageFactory`, passing the address, request type, and header lines as parameters. The application can then send the request by calling the `sendRequest()` operation on the `SipProvider` with the `Request` object as parameter. The resulting Java code looks like this:

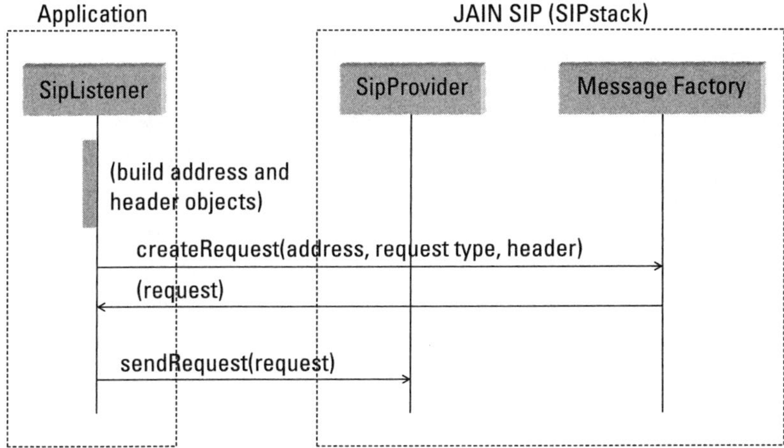

Figure 4.8   Sending a SIP request.

```
requestInstance=messageFactoryInstance.
createRequest(
     requestUriInstance, "INVITE",
     callIdHeaderInstance,
     cSeqHeaderInstance, fromHeaderInstance,
     toHeaderInstance, viaHeaders,
     maxForwardsHeaderInstance,
     contactHeaderInstance);
sipProviderInstance.sendRequest(requestInstance);
```

In this code snippet we assume that the application has previously created and configured the objects that specify the request address `requestUri-Instance` and header lines `callIdHeaderInstance`, `cSeqHeaderIn-stance`, `fromHeaderInstance`, `toHeaderInstance`, `viaHeaders`, `maxForwardsHeaderInstance`, `contactHeaderInstance`.

The application can build and send SIP responses in a similar way, using the `createResponse()` and `sendResponse()` operations instead.

4.3.5   Keeping the SIP Transaction State

In the example of Section 4.3.4, the application sets the session- and trans-action-related `CSeq:` and `Call-Id:` headers itself before sending the mes-sage. Most SIP requests expect matching responses, and the application will have to keep track of the state of the transaction (request-response pair) so it can relate an incoming response to a sent request or a sent response to an incoming request.

The JSIP API lets the application keep the transaction state for an outgoing request through the `ClientTransaction` object, shown in Figure 4.9.

The application creates a request object in the same way as in Figure 4.8. In this case, however, the application does not send the request directly through the `SipProvider` interface but asks the `SipProvider` to create a `ClientTransaction` object. The application then sends the request by calling the `sendRequest()` operation on the `ClientTransaction` object, which records the state of the transaction.

In a similar way, an application can ask the `SipProvider` to create a `ServerTransaction` object for an incoming request and send the response through this object so as to correlate it with the request.

In Section 3.3.6 we've seen that SIP transactions can form a dialog. Sometimes an UA will need to send a request as part of a specific dialog, such as when it has to cancel an outstanding request (CANCEL) or end a SIP session in progress (BYE). Figure 4.10 shows how an application can relate a request to an ongoing dialog using the `Dialog` interface.

To relate a request to a dialog, the application uses the `Dialog` interface as the request factory. The `Dialog` object automatically relates the created request to its dialog. The application then requests new `ClientTransaction` and sends the request through the `Dialog`, so the `Dialog` keeps track of the transaction as part of the dialog.

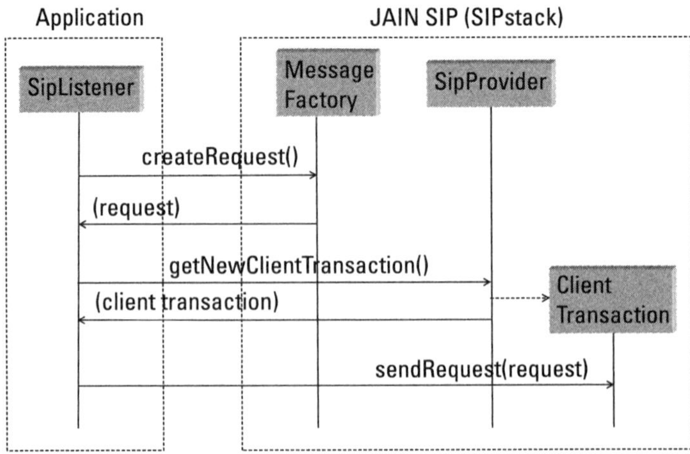

Figure 4.9 Keeping transaction state for an outgoing SIP request.

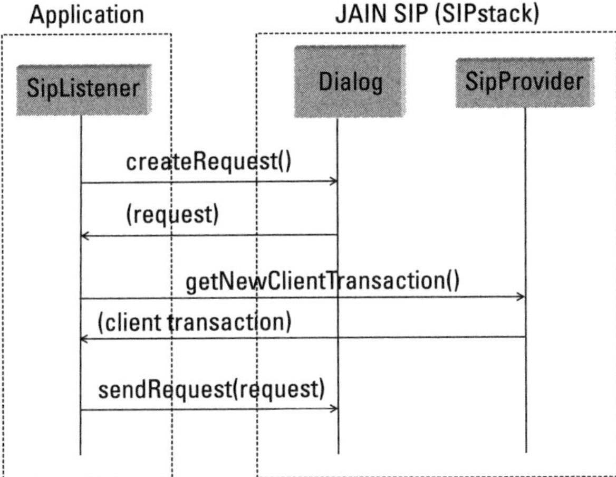

Figure 4.10   Sending a request within an ongoing SIP dialog.

4.3.6 SDP API

In Section 3.3.5 we've seen that SIP messages often carry SDP payloads to describe and negotiate session parameters such as media endpoints and codecs. Java also provides a package to parse and assemble SDP documents that can travel in the body of a SIP message, called the SDP API [6]. The SDP API is a simple API with two main interfaces:

- `SdpFactory` allows the application to create a new SDP document. Like `SipFactory`, it uses the Singleton pattern to ensure that there is exactly one `SipFactory` instance.
- `SessionDescription` represents an SDP document that can be put in the body of a SIP message. Remember that an SDP document is no more than a series of "attribute=value" pairs where single letters define the attribute type ("v" for version, "o" for origin, "n" for session name, "m" for media descriptors, "a" for media attributes, and so on; see Section 3.3.5).

A `SessionDescription` object contains objects that represent the SDP lines. Table 4.2 defines the classes that correspond to the main SDP lines.

An application can create a new empty `SessionDescription` object by calling the `createSessionDescription()` operation on the

Table 4.2
SDP API line classes

Class	SDP Line
Version	SDP version ("v=" line)
Origin	Session owner ("o=" line)
SessionName	Session name ("n=" line)
Connection	Connection address ("c=" line)
MediaDescription	Media description and attribute ("m=" and "a=" lines)

`SdpFactory`. The `SdpFactory` also creates objects of the classes shown in the left column of Table 4.2 when called with the `createVersion()`, `createOrigin()`, or `createMediaDescription()` operations, which carry the parameters to define the corresponding line.

The application can then set the lines in the SDP document using `set()` operations for each of the classes in Table 4.2 on the `SessionDescription` object, such as `setConnection(Connection c)`. The application can get header lines from a `SessionDescription` object with `get()` operations for each line, such as `Connection getConnection()`.

4.3.7 SDP API: An Example

The best way to grasp how to use the SDP API is by an example. The Java code snippet shown next creates a session description that corresponds to the SDP example of Table 3.5 in Section 3.3.5. This code omits the creation of the a, i, and e lines in Table 3.5 so as to keep the example concise:

```
/*Bootstrap SdpFactory using the Singleton
pattern*/
SdpFactory sdpFactoryInstance = SdpFactory.
getInstance();
/*Create an empty session description*/
SessionDescription sessionDescriptionInstance =
                    sdpFactoryInstance.
createSessionDescription();
```

```
/*Create and set v line*/
Version v = sdpFactoryInstance.createVersion(0);
sessionDescriptionInstance.setVersion(v);
/*Create and set o line*/
Origin o = sdpFactoryInstance.
createOrigin("Zuidweg", 3321578898, 1,"IN",
            "IP4", "194.140.170.11");
sessionDescriptionInstance.setOrigin(o);
/*Create and set s line*/
SessionName sessionNameInstance =
sdpFactoryInstance.createSessionName("SIP and SDP
course");
sessionDescriptionInstance.setSessionName(s);
/*Create and set c line*/
Connection c = sdpFactoryInstance.
createConnection("IN", "IPv4",
                "194.140.170.15/127");
sessionDescriptionInstance.setConnection(s);
/*Create and set m line*/
MediaDescription m =
        sdpFactoryInstance.
createMediaDescription("audio", 49170, 1,
                       "RTP/AVP", 0);
sessionDescriptionInstance.setMediaDescription (m);
```

First, this example shows how to get an instance of the `SdpFactory`, which creates all other objects. The function `SdpFactory.getInstance()` in the second line shows that `SdpFactory` uses the Singleton pattern to obtain its single instance, just like the `SipFactory` in Section 4.3.3. After obtaining an instance of the `SdpFactory`, the code creates an empty `SessionDescription` object, and then creates and inserts each line of the session description.

Each part of the session description is represented by object created with the `SdpFactory`. For example, the c line in the session description of Table 3.5 is represented by a `Connection` object, created by calling `createConnection` with parameters "`IN`," "`IPv4`," and the IP address `194.140.170.15/127`.

The `set` operations aggregate the description objects into an instance of the `SessionDescription`; for example, `sessionDescripionInstance`.

`setConnection(c)` includes the `Connection` object c into the `session-DescripionInstance` object.

SDP allows descriptions to consist of several lines (e.g., media descriptions can consist of more than one m line and can also contain a lines; see Table 3.5). The SDP API supports composite descriptions by letting the programmer arrange the corresponding objects into vectors. We won't go into this in more detail, as [6] explains this.

When the `sessionDescriptionInstance` captures the complete session description after executing the code, the programmer can generate a plain text SDP description from it by calling the following function:

```
sessionDescriptionInstance.toString().getBytes();
```

4.4 SIP Servlets

Java's SIP Servlets API [7] provides utilities for building SIP applications at a higher level of abstraction than the JSIP API. A SIP servlet is an application component that does SIP signaling and does not maintain a state. In other words, a SIP servlet can send SIP messages, respond to SIP messages, and do additional processing, but it does not keep persistent data for the SIP dialog or the application it belongs to.

A SIP servlet forms part of an application that may consist of additional objects and resources such as data or content. An application may use more than one SIP servlet, and a SIP servlet may be reused in more than one application.

4.4.1 SIP Servlet Container

Strictly speaking, SIP servlets provide more than just an API. Like HTTP servlets, SIP servlets also provide a container model for servlets to run in.

A container is an object that provides an execution environment for the objects it contains. It also provides general functions to manage the life cycle (creation, initialization, destruction) and other noncritical aspects of its contained objects. This probably sounds somewhat abstract, but in the course of this section we'll see more details of the facilities the container provides.

Figure 4.11 shows a general view of the SIP servlets model in the context of an IMS AS. Figure 4.11 illustrates how SIP servlets run in a SIP servlet container and process SIP messages as calls to Java objects.

Figure 4.11 shows the default situation where the SIP servlets and applications run in the same AS on the same virtual machine. An application may

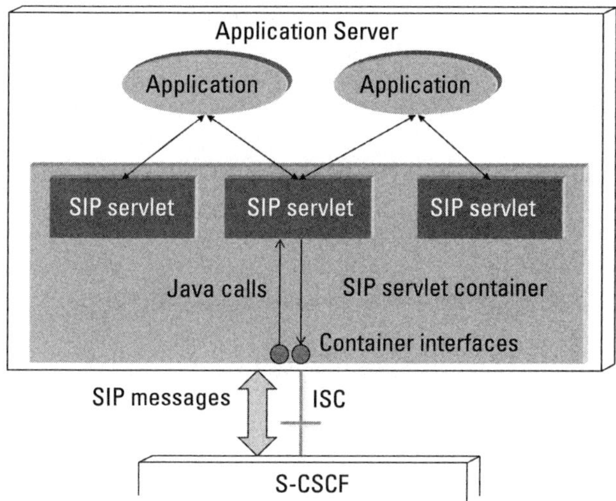

Figure 4.11　SIP servlets concept.

also have a distributed deployment (i.e., its SIP servlets and components may run on different virtual machines).[7]

A SIP servlet container encapsulates a SIP stack that parses and assembles SIP messages, and manages SIP transactions and dialogs. For simplicity, Figure 4.11 does not show the encapsulated SIP stack.

One purpose of the container is to manage the life cycle of the SIP servlets it contains. When an application needs a SIP servlet instance, the SIP servlet container loads the SIP servlet class, creates an instance, and initiates the newly created SIP servlet by calling the `init()` operation on it. Once created, the container can route SIP events to the SIP servlet, and the application can use the SIP servlet to send SIP messages. When the SIP servlet becomes obsolete, the container can disable it by calling the `destroy()` operation on it. The Java garbage collector[8] then frees up the server's memory that was used by the SIP servlet.

---

[7] Distributed applications need additional Java APIs to ensure that objects running on different virtual machines can interact. Java remote method invocation (RMI) is such an API. In this book, we won't explain distributed Java applications in more detail, but Chapter 5 explains the principle of web services, SOAP, and REST, which are used to build distributed applications.

[8] Java's run time system includes a garbage collector. The role of the garbage collector is to reclaim computer memory occupied by objects that are no longer being used by the program. An object can be considered "dead" if no other object has any reference to it. The garbage collector frees the programmer from having to keep track of an object's use and from having to clean up obsolete objects manually. Without the garbage collector, memory can easily fill up with unused objects,

An application can receive notifications about the life cycle of its SIP servlets by implementing the `ServiceLifeCycleListener` interface and registering it to the container.

4.4.2 SIP Servlets and HTTP Servlets

If you've heard about HTTP servlets you probably wonder how they relate to SIP servlets. Indeed the name *servlet* is not a coincidence and indicates that SIP servlets and HTTP servlets inherit from the same generic `Servlet` interface and can run in the same container. SIP servlets use the same directory structure as HTTP servlets, and they can be deployed together as application components.

However, SIP servlets also differ in many ways from HTTP servlets because of the variations between the SIP and HTTP protocols. Most importantly, HTTP servlets always run on HTTP servers, whereas SIP servlets can run on clients (UAC), servers (UAS), and intermediate nodes (proxy servers and redirect servers). This implies that whereas HTTP servlets only reply to requests, SIP servlets must also be able to send requests and receive responses.

But as SIP servlets and HTTP servlets share the same base interface and can run in the same container, the programmer has a consistent view of both and can combine them to implement rich communications applications. One example is the call back when available service we saw in Section 3.4.3, which can be implemented by using a SIP servlet for SIP signaling and a HTTP servlet for handling the HTTP requests.

4.4.3 Processing SIP events

SIP servlets allow applications to use SIP signaling on a higher level of abstraction than the JSIP API. The SIP servlet container frees the SIP servlet programmer from having to assemble SIP messages and keep track of the state of SIP transactions and dialogs. Using SIP servlets, the application does not have to worry about setting the right `To:`, `From:`, `Call-Id:`, `CSeq:`, and `Via:` headers (see Section 3.3.4) and can concentrate on higher-level tasks such as completing or proxying the call, translating the destination address, or negotiating media parameters with SDP. The SIP servlet container encapsulates the low-level aspects of SIP signaling, such as maintaining the consistency of the headers and correlating responses to requests.

---

which can eventually lead the application or (worse) the system to crash. Those who have programmed in C will be familiar with the problem.

Table 4.3
SIP Servlet Notifications

SIP Request	SIP Servlet Notification
REGISTER	doRegister()
INVITE	doInvite()
ACK	doAck()
BYE	doBye()
CANCEL	doCancel()
OPTIONS	doOptions()

SIP servlets use the Observer pattern to register to SIP events and to receive them. However, SIP servlets implement this pattern in a different way than the JSIP API.

When the SIP servlet container receives a SIP request or response from the SIP stack, it creates a corresponding `SipServletRequest` or `SipServletResponse` object. It then notifies the listening SIP servlets by calling the `service()` operation on the general `Servlet` interface of the SIP servlets with the `SipServletRequest` or `SipServletResponse` object as parameter. The SIP servlet dispatches this call to an internal `doRequest()` or `doResponse()` operation depending on the `SipServletRequest` or `SipServletResponse` object provided by the container. SIP servlets do this internally and the SIP application programmer only needs to implement the handling of each specific `doRequest()` or `doResponse()` event. Each SIP request has a corresponding `doRequest()` operation, as shown in Table 4.3.

Likewise, each SIP response (see Section 3.3.4) has a corresponding `doResponse()` operation: `doSuccessResponse()` for SIP 2xx responses, `doProvisionalResponse()` for 1xx responses, `doRedirectResponse()` for 3xx, and `doErrorresponse()` for 4xx, 5xx, and 6XX responses.

By default the SIP servlet class implements the `doRequest()` and `doResponse()` operations as follows:

- All `doResponse()` operations and `doAck()` and `doCancel()` operations do nothing.
- `doInvite()`, `doRegister()`, `doOptions()`, and other requests return a 501 ("not implemented") error code if the request is an initial request, or else they do nothing.

The application programmer defines the desired processing for each request or response by overriding[9] the implementation of these `doRequest()` and `doResponse()` operations.

A `doRequest()` or `doResponse()` operation can handle the request or response synchronously, but it can also store the request in a `SipSession` object for later, asynchronous processing.

The SIP servlet container relies on the SIP servlet to handle incoming requests and responses. It does not send default responses and does not complete dialogs by itself, except when receiving a `CANCEL` request. In this case, the container automatically returns a `200 OK` response. It calls `doCancel()` on the corresponding servlet, but the servlet does not need to respond explicitly. Also, a SIP servlet container does not notify a servlet of an incoming `ACK` because this only completes a three-way handshake and is generally not of interest to the application.

In the next section, we'll see how the SIP servlet container lets the SIP servlet send a request or response as part of a particular SIP transaction or dialog.

4.4.4 SIP Sessions and Dialogs

The SIP servlets API provides two interfaces that help the application and the SIP servlet keep the state of transactions and dialogs:

- A `SipApplicationSession` provides a SIP servlet with an interface that represents the application it belongs to. A SIP servlet creates a new `SipApplicationSession` instance when an incoming SIP event triggers the execution of an application or when an application invokes the SIP servlet to start SIP signaling.
- A `SipSession` represents an exchange of SIP messages between two end points. A `SipSession` roughly represents a SIP dialog, but SIP servlets must also bind requests that do not belong to a dialog to a `SipSession`. In other words, all SIP requests must belong to a `Sip-Session`, whether or not they belong to a dialog. For those requests sent as part of a dialog, the `SipSession` keeps the state of the dialog, or else it just encapsulates a single transaction or even a single message if it does not belong to any transaction or dialog.

---

[9] When a class inherits from another class, the programmer can override (i.e., reprogram) the operations that were inherited from the superclass, giving them extended or new functionality but without changing the class's interface.

A `SipApplicationSession` object contains `SipSession` objects that represent the active SIP sessions that the application maintains at any point in time. An application can engage in more than one SIP session at a time (e.g., if it acts as a back-to-back UA; see Section 3.4.3), so `SipApplicationSession` and `SipSession` objects have a one-to-many relationship.

The SIP servlet container creates new a `SipSession` object when an initial SIP request (not yet belonging to any SIP session) comes in from the network or when a SIP servlet sends an initial outgoing SIP request. The SIP servlet container automatically relates subsequent responses and requests to the appropriate `SipSession` object.

`SipSession` objects can have four states, as shown in Table 4.4.

If the `SipSession` receives a response indicating failure (3xx, 4xx, 5xx, 6xx) in the `INITIAL` or `EARLY` state, it will set the state back to `INITIAL`.

4.4.5 Creating and Sending SIP Messages

All incoming and outgoing `SipServletRequest` and `SipServletResponse` must belong to a `SipSession`, and each `SipSession` belongs to a `SipApplicationSession`, as we've seen in the previous section. In this section, we'll see how a SIP servlet creates an initial request, responds to incoming requests, and sends subsequent requests.

To send an initial SIP request, the SIP servlet must first obtain a new `SipApplicationSession`. The SIP servlet container provides the `SipFactory` interface for this purpose and for creating initial SIP requests that start a new `SipSession`. A SIP servlet creates all subsequent SIP requests through the `SipSession` object they belong to and creates responses directly on the incoming `SipSession`. Figure 4.12 shows the relationship between the different classes.

Table 4.4

`SipSession` States

State	Semantics
INITIAL	Initial request sent
EARLY	Early request (1xx) received
CONFIRMED	Success response (2xx) received
TERMINATED	Error response (4xx, 5xx, 6xx) or termination request (BYE, CANCEL) received

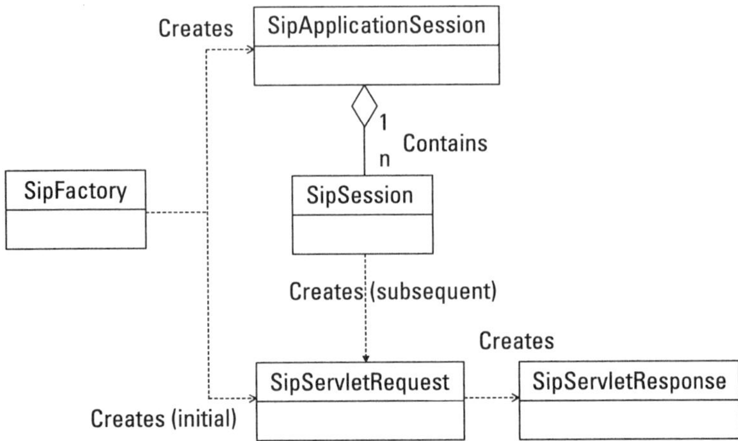

Figure 4.12   Creation of SIP sessions and messages.

SIP servlets can send SIP requests and responses by first creating a
`SipServletRequest` or `SipServletResponse` object with the necessary parameters and then calling the `send()` operation on it. Creating and
sending an initial SIP request looks like this, assuming that the SIP servlet has
first obtained a new `SipApplicationSession` in the appSesion variable and that it has previously created the `fromAddress` and `toAddress`
objects for the request:

```
initialRequest = sipFactoryInstance.createRequest(
            appSession, "INVITE", fromAddress,
            toAddress);
initialRequest.send();
```

This code automatically creates a new `SipSession` object representing
the newly initiated SIP message exchange under the `SipApplicationSes-`
`sion` instance appSession. If the SIP servlet has to send follow-up requests,
it will send them through the corresponding `SipSession` object:

```
cancelRequest = sipSessionInstance.createCancel();
cancelRequest.send();
```

This example also shows that a SIP servlet can directly create CANCEL
and BYE messages with the parameterless `createCancel()` and `create-`
`Bye()` messages. The servlet does not need to provide the `fromAddress`
and `toAddress` again in this follow-up request because the `SipSession`
remembers them and fills them automatically.

SIP servlets make responding to incoming messages equally straightforward. Responses are sent through the `SipServletRequest` they are responding to, and the container automatically correlates the response with the request. Assuming that `incomingRequest` holds the incoming `SipServletRequest` object, the servlet can generate a response as follows:

```
response = incomingRequest.
createSipServletResponse();
response.send();
```

By default, the `createSipServletResponse()` operation creates a success response that matches the request. The SIP servlet can change this by redefining the headers in the `SipServletResponse` object before calling `send()` on it in the second line. In the following section, we'll see how SIP servlets can manipulate `SipServletRequest` and `SipServletResponse` objects.

4.4.6 Processing SIP Requests and Responses

The `SipServletRequest` and `SipServletResponse` classes provide operations that allow an application to get and set header lines. Table 4.5 provides an overview of the operations available to manipulate a request or response header object.

To build address headers, the SIP servlet API provides an `Address` class that has a similar structure as in the JSIP API. Many operations in the SIP servlet API accept addresses as `String` (weakly typed) or as `Address` (strongly typed).

SIP servlets can get and set the content in the body of SIP messages just like HTTP servlets, but in a different way. HTTP messages use a streaming model for accessing the message body because the payload can have substantial size (e.g., a HTML page with images or videos). As SIP messages tend to have much smaller payloads (most often an SDP document of a few lines), SIP servlets use the same mechanism as the Java Mail API [8] to access the body of a SIP message.

An application can extract the content of a `SipServletRequest` or `SipServletResponse` through its `getContent()` operation. This overloaded[10] operation can return the content as a Java object or as raw bytes.

---

[10] An overloaded operation is an operation that has various implementations with different parameters and result types. Examples include `int calculateSum (int x, int y)` and `float calculateSum (float x, float y, float z)`. Formally these are different operations

Table 4.5
Operations to Get and Set Message Headers

Operation	Effect
```void setHeader(String headerName, String headerValue)```	If the message already contains a header line of type `headerName`, this operation overwrites the header with `headerValue`. Otherwise, it creates a new header line with the specified value.
```void addHeader(String headerName, String headerValue)```	Creates a new header line of type `headerName` with the value `headerValue`.
`getHeader(String headerName)`	Returns the first header of type `headerName`.
`listIterator getHeaders()`	Returns an iterator[1] over the list of headers in the message. When called with the `headerName` argument, this operation returns an iterator over the list of headers of the specified type.

[1] An iterator is an object that represents a list and provides sequential access to the list elements. It is described as a design pattern in [2].

An application can insert content into `SipServletRequest` or `SipServletResponse` through its `setContent(Object obj, String contentType)` operation, passing the content as Java object and providing a text description of the content type.

4.4.7 Composing SIP Applications

A SIP servlet container can have more than one application listening for incoming SIP messages through different SIP servlet instantiations. This immediately raises the question how and in which order the container must notify the different applications of an incoming SIP message. How the container dispatches SIP messages to listening applications can influence how these applications interact or (worse) interfere.

To avoid complex and unpredictable interference between applications, the SIP servlet container treats applications as strictly separate entities that

because they have different signatures (input and output parameters and types), but they share the same name.

must be triggered in sequence, one after the other. When the first application has processed an incoming SIP message and produced an outgoing SIP message, the container feeds that message into the next application and so forth.

Though this reduces the complexity of the interactions between applications on the same application server, the container will still need to determine the order in which to notify its SIP servlets. This order can be significant.

Imagine the case of a user who has deployed two applications, each implemented by a SIP servlet. One servlet provides a calling line identifier feature (displaying the identity of the calling party to the user) and the other servlet screens the incoming call (using a black list, time of day, or other criteria). When the SIP servlet container receives an incoming SIP `INVITE` message, which servlet should it notify first?

In this case, the container must trigger the call screening servlet first because it should not allow the calling line identifier servlet to alert the user if the call was to be barred by the call screening servlet.

To know in which order to notify its SIP servlets of incoming SIP events, the SIP servlet container consults an application router. A service deployer programs the application router to define the order in which the container must notify its SIP servlets of incoming SIP messages. Figure 4.13 illustrates the role of the application router.

Although the SIP servlets specification defines the application router interface, it does not specify how to implement it. It is up to the application designer or deployer to provide the implementation of this component.

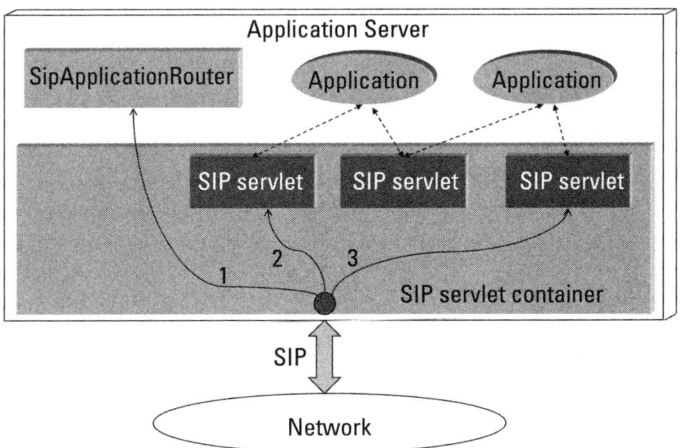

Figure 4.13 The application router.

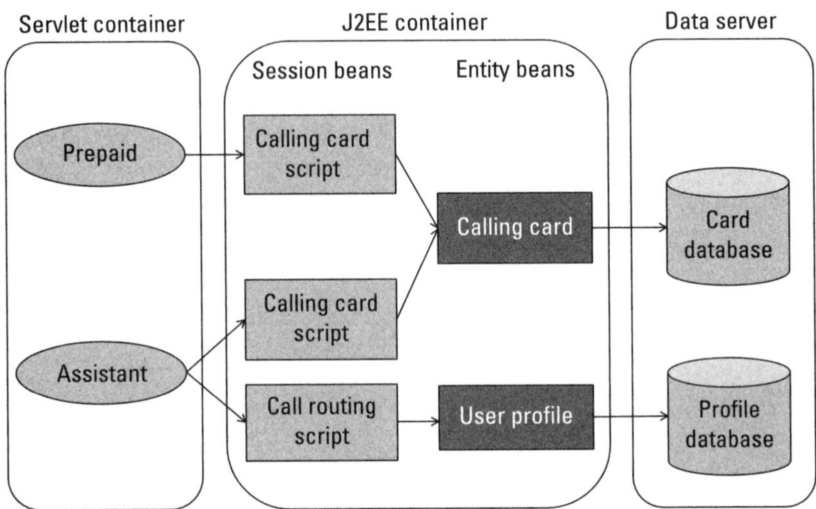

Figure 4.14   Using SIP servlets in JEE applications.

4.4.8 Building JEE Applications with SIP Servlets

The SIP servlets model makes no specific assumptions on the applications that use SIP servlets. An application can consist of just a SIP servlet but can also contain additional classes and resources such as content (media files). SIP servlets can even form part of large-scale JEE applications. Remember that JEE provides the Java run time environment and API for large-scale enterprise applications (see Section 4.2.3).

In this section, we'll see an example of how to build a SIP application from SIP servlets and JEE components. Figure 4.14 provides an example of an AS that has two deployed applications, a prepaid calling application and a call assistant application.

One SIP servlet represents the prepaid call application, and another SIP servlet represents the call assistant application. The SIP servlets handle only SIP signaling and delegate all other application processing to JEE components, called *Java beans*.[11]

There are two types of JEE components:

- *Session beans* define an application's control flow. Session beans have a limited life span. They are created to accomplish a task and then cease to exist.

---

[11] Java beans are JEE components that run in a JEE container.

- *Entity beans* represent persistent data that can reside on external databases. Entity beans typically have a longer life span and can be reused among session beans.

The applications in this example employ two types of session beans:

- The *calling card script* session beans execute the calling card application logic. When a subscriber wants to make a call, this session bean first checks the calling card credit and the call destination to determine whether the subscriber has sufficient credit to complete the call. It then instructs the prepaid SIP servlet to send a SIP `INVITE` request to the destination to set up the call and proceeds by updating the subscriber's credit as the call progresses in a very similar way as described in Section 2.9.1 (prepaid calling with CAMEL).
- The *call routing script* session beans execute application logic for call routing. When the call assistant SIP servlet receives an incoming `doInvite()` call, it invokes the call routing script session bean. The call routing script session fetches the user profile entity bean for the subscriber and instructs the SIP servlet to complete the call according to the user's preferences.

The applications use two types of entity bean:

- The *calling card* entity beans store the prepaid credit for a particular subscriber.
- The *user profile* entity beans store preferences for a particular subscriber, including how and where the subscriber would like to be reached. For example, this entity bean can record that calls to this subscriber have to be routed to his or her office phone between 9 AM and 5 PM, to his or her mobile phone between 5 PM and 10 PM, and to voice mail between 10 PM and 9 AM.

The prepaid call application consists of the prepaid SIP servlet, the calling card script session bean, and the calling card entity bean. The call assistant application uses the assistant SIP servlet and two session beans: the call routing script and the calling card script. It also involves two entity beans: the calling card entity bean and the user profile entity bean. This demonstrates that a single application can involve more than one session bean and entity bean. An application can also use more than one SIP servlet.

Note that the session and entity beans do not need any knowledge of SIP. The SIP servlets encapsulate all SIP signaling and have only application-specific interactions with the Java beans (e.g., check the prepaid credit, get the call completion address).

Application designers can build SIP and IMS applications of any complexity combining SIP servlets with JEE components. In the next section, we'll discuss a Java API and execution environment that provides an even higher-level abstraction for programming communications applications and that can also be combined with JEE.

4.5   JAIN SLEE

SIP servlets and the JSIP API allow a programmer to build SIP applications at two different levels of abstraction. Although SIP servlets free the programmer from SIP message handling chores such as relating SIP responses to SIP replies and keeping transaction and dialog state, SIP servlet programming still requires substantial knowledge of the SIP protocol and its messages.

JAIN SLEE [9] offers an execution model and API with a higher level of abstraction, allowing a programmer to build applications and services that use a variety of communications protocols without requiring detailed knowledge of the protocol structures. JAIN SLEE enables carrier-grade[12] applications that satisfy high-performance, availability, and security standards.

The price to pay is that JAIN SLEE has a more complex execution model than the JSIP API or SIP servlets. In a way, JAIN SLEE trades protocol complexity for complexity of the execution model. However, JAIN SLEE does facilitate the reuse of preprogrammed components, which enables platform providers to build almost plug-and-play application creation environments for nonexperts.

In line with the purpose of JAIN SLEE we'll provide an overview of JAIN SLEE in this section without diving too deeply into specific Java code.

4.5.1   JAIN SLEE Overview

JAIN SLEE applications run in a container just like SIP servlets, but the container offers more general execution support such as data persistence, availability, and security. Like SIP servlets, JAIN SLEE is event driven and uses the Observer pattern to notify applications of incoming signaling messages.

JAIN SLEE applications can use different signaling protocols through resource adapters that encapsulate the protocol stack for each protocol. Through

---

[12] Carrier-grade systems must be able to handle more than 1,000 transactions per second with a latency of less than 0,1s and to have 99,999% ("five nines") uptime.

its protocol adapters, JAIN SLEE provides applications with an abstract view of the underlying network. JAIN SLEE has resource adaptors for SIP, INAP, and other protocols. JAIN SLEE applications are built from service building blocks (SBBs) that contain reusable service logic. SBBs run in a JAIN SLEE container that provides them with a carrier grade execution environment and manage their life cycle. Figure 4.15 illustrates the general JAIN SLEE architecture.

Figure 4.15 shows that the JAIN SLEE container provides applications with a wide range of utilities, including timer, tracer, alarm, and management services. In the following sections, we'll take a closer look at some of the JAIN SLEE features, and we'll discuss the role of the JAIN SLEE resource adapters, in particular Java call control (JCC).

4.5.2 Service Building Blocks

Service building blocks (SBBs) are the components from which JAIN SLEE applications are built. JAIN SLEE refers to the class defining a SBB as the SBB component, and calls an instantiation of an SBB class an SBB entity.

SBB entities can create other SBB entities at run time and can define an order to propagate events to children SBB entities. An SBB entity tree represents this relationship between SBBs and defines how the container should route events to SBBs. This differs from SIP servlets where the container has to consult the application router to determine how to route events to SIP servlets (see Section 4.4.7). Figure 4.16 shows an example of an SBB entity tree. In Figure 4.16, SBB X has created two entities of SBB Y, and the second SBB Y entity has created another SBB Y entity and an SBB Z entity. The container

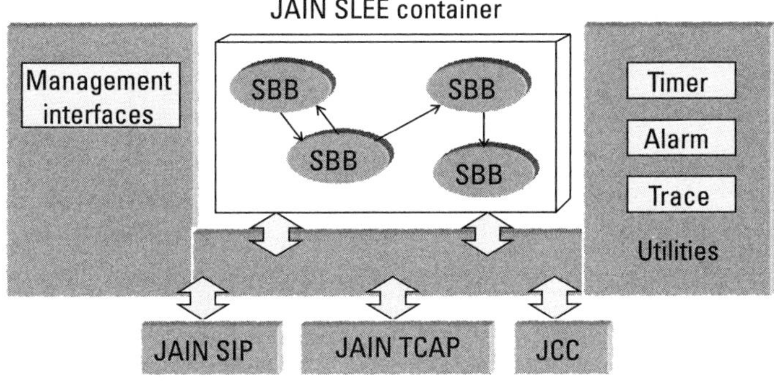

Figure 4.15 JAIN SLEE overview.

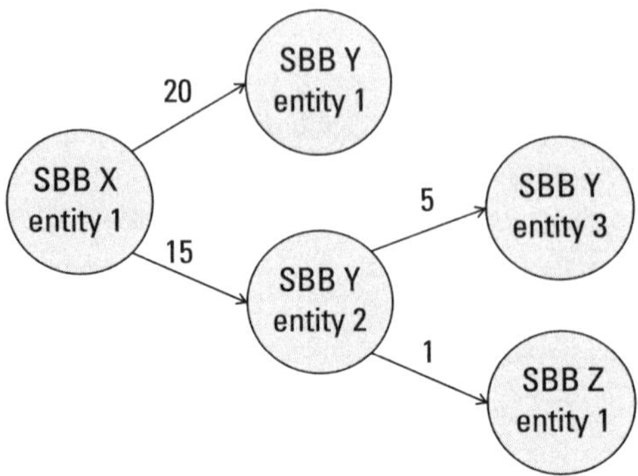

Figure 4.16   Example SBB entity tree.

propagates events from SBB X entities to SBB Y entities with default priority 20, but in the case of the second SBB Y entity, this priority has been overridden to 15. The container propagates events from the SBB Y entity to the child SBB Y entity with priority 5, to the SBB Z entity with priority 1.

An SBB component has two interfaces:

- The *standard interface* provides the operations for handling incoming events, generating outgoing events, managing its SBB entity tree, and managing persistent data.
- The *local interface* provides operations available to other SBBs for application-specific interactions. Figure 4.17 gives a schematic view of the interfaces an SBB exhibits.

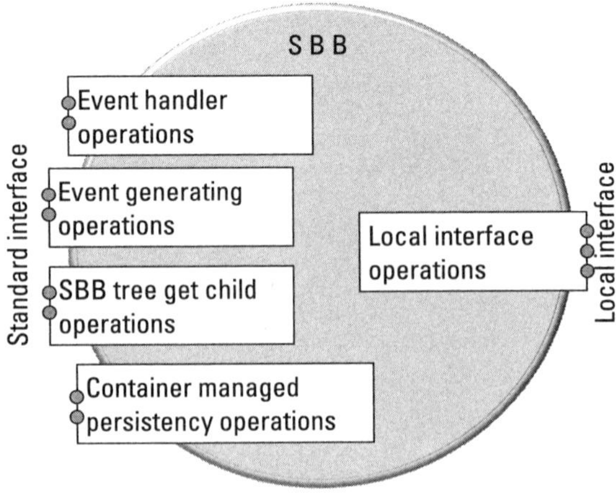

Figure 4.17   SBB component interfaces.

JAIN SLEE facilitates the creation of SBBs by generating an abstract class from an SBB interface specification. The programmer implements the concrete class implementing the operations to handle events, store persistent data, and so on.

4.5.3   Activities and Events

JAIN SLEE has a more extensive and more generally applicable event model than the JSIP API or SIP servlets. In JAIN SLEE, events belong to activities. Activities are abstractions of event-generating processes such as a voice call. SBB entities can subscribe to the events from an activity through its `Activity-Context` interface. Figure 4.18 illustrates how SBB receive events from activities.

Activities are tightly related to resource adapters because resource adapters represent event-generating processes such as SIP dialogs. In the next section, we'll study a specific resource adapter and its activity model.

4.5.4   JCC Resource Adapter

JAIN SLEE uses resource adapters to provide abstract event driven views of network protocols and other resources. The JSIP API is one of the resource adapters included in JAIN SLEE. As we've seen in Section 4.2, the JSIP API provides rather low-level access to the SIP protocol stack.

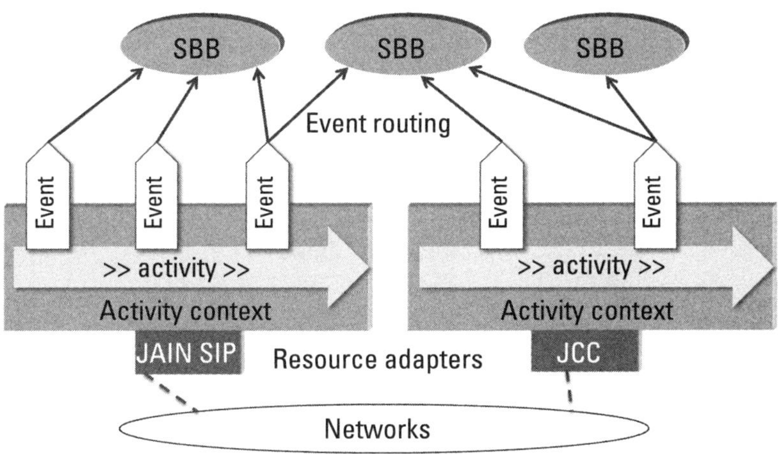

Figure 4.18   JAIN SLEE event model.

JAIN SLEE also has resource adapters that provide more abstract activities and events, such as Java call processing and (JCP), Java call control (JCC), and Java coordination and transactions (JCAT). JCP, JCC, and JCAT provide activity models for call setup very similar to the IN BCP we saw in Section 2.4.2. JCP, JCC, and JCAT have different granularities, with JCP providing the least and JCAT the most detailed call model. In this section, we'll focus on JCC [10], which provides an intermediate call model similar to the IN BCP.

The JCC API has four main interfaces:

- `JccPeer` encapsulates the implementation of JCC and has a role similar to the `SipStack` in the JSIP API. An application can initialize the JCC software and obtain an instance of the `JccProvider` through this interface.
- `JccProvider` lets an application create calls and register for call-related events. Its role is similar to the role of the `SipProvider` interface of the JSIP API.
- `JccCall` represents an incoming or outgoing call. This interface allows an application to `JccConnection` objects when it receives a request to connect parties in a call.
- `JccConnection` represents a connection within a call to an endpoint address, which is represented by a `JccAddress` object. A call can have zero or more connections. A two-party call normally has two connections, one for each party's endpoint address.

JCC also uses the Observer pattern. Applications can request to receive events from `JCCProvider`, `JccCall` or `JccConnection` activities through their `JccProviderListener`, `JccCallListener`, or `JccConnectionListener` interfaces. Figure 4.19 shows the structure of the JCC API.

As we saw in Section 4.5.3, JAIN SLEE structures events by grouping them into activities. The `JccCall` object is such an event-generating activity, which represents the states of a call. Figure 4.20 shows the states of the JCC call model. A `JccCall` can send an event to its observers every time the call changes state. The event carries relevant data such as the calling party and called party address.

You may have observed that the JCC call model bears certain resemblance to the IN BCSMs of Section 2.5.3. JCC provides a Java programming environment for IN-like call control.

Now let's look at an example of a call forwarding application, programmed with the JCC API. In this example, we assume that the SBB has

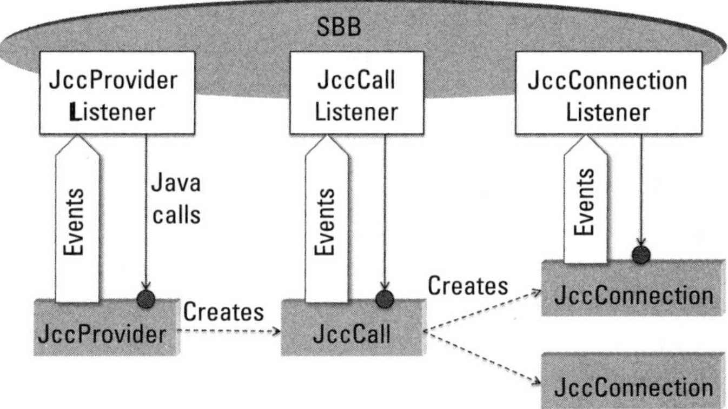

Figure 4.19 JCC structure.

previously registered its `JccConnectionListener` to receive connection related events.

Figure 4.21 shows the following message flow for a basic call forwarding application:

1. Upon receiving a call (through encapsulated SIP or TCAP signaling), the `JccProvider` creates a corresponding `JccCall` object to represent the incoming call.

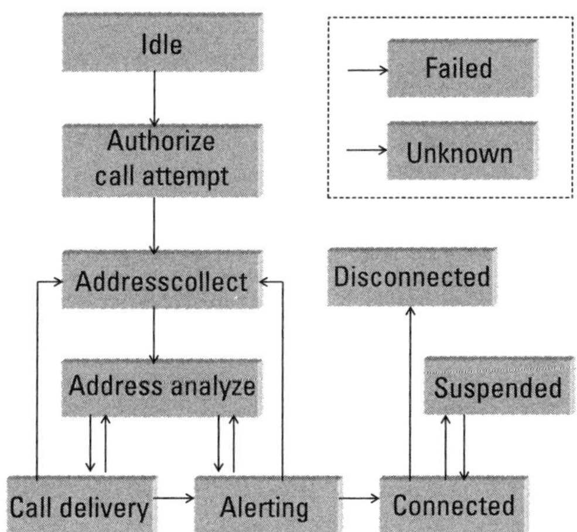

Figure 4.20 JCC call event model.

2. The `JccCall` object creates `JccConnection` objects for the originating and terminating endpoints. For simplicity, Figure 4.21 only shows only the `JccConnection` interface for the destination address of the call.

3. The `JccConnection` notifies the registered `JccConnectionListener` of its creation.

4. The application retrieves the destination address for the call by calling the `getAddress()` operation on the `JccConnection`. The `JccConnection` returns the destination address for the call, 696848485.

5. The SBB instructs the `JccCall` to reroute the call to number 696014239 with the `routeCall()` operation.

6. The `JccCall` creates a new `JccConnection` object with associated address 696014239. The encapsulated JCC software takes care of setting up an actual connection to this address using SIP or SS7 signaling.

7. The `JccCall` tells the old `JccConnection` to kill itself by calling the `release()` operation on it. As a result, the encapsulated JCC software releases the connection address 696848485.

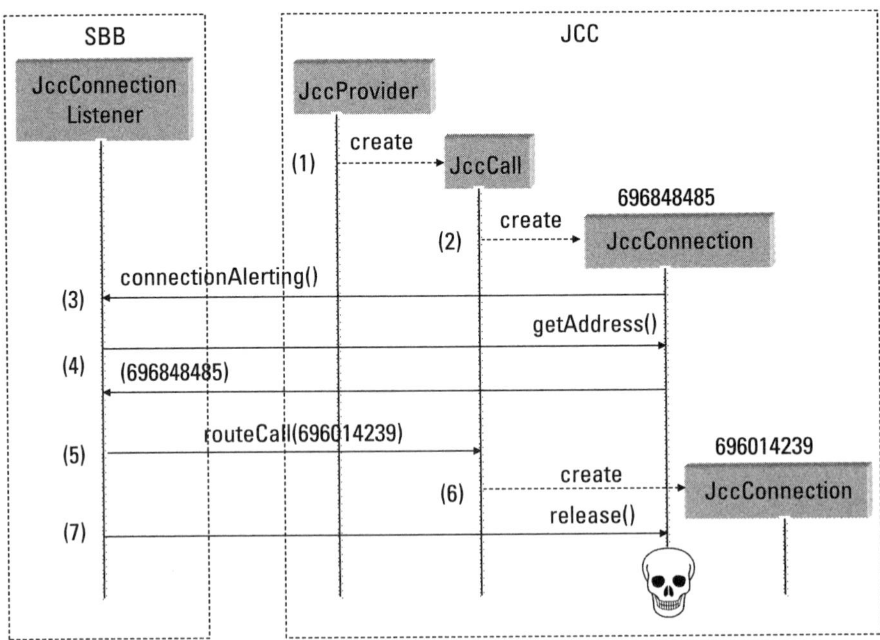

Figure 4.21   Call forwarding with JCC.

JCC allows applications to control outgoing calls in a similar way as shown in the previous example, using the `JccProvider`, `JccCall`, and `JccConnection` interfaces.

Different resource adapters allow an application designer to build JAIN SLEE applications that interact with different protocols. For example, with JAIN SLEE it is possible to implement an application that can handle both IN and IMS calls. In the next section, we'll see that it is also possible to connect JAIN SLEE with enterprise applications.

4.5.5 Combining JAIN SLEE with JEE and JEE

JAIN SLEE and JEE both provide container-based execution environments, but with a different purpose. As we've seen in this section, JAIN SLEE enables carrier-grade communications applications for different types of network including IMS and circuit switched networks. JEE, on the other hand, is used to program enterprise applications that depend on large amounts of data. JAIN SLEE is for carrier-grade telecommunications services, whereas JEE is for enterprise-scale information processing. But that doesn't mean that they cannot be combined.

Although the JAIN SLEE component model differs from the JEE component model, SBBs in a JAIN SLEE container can connect to Java beans running in a JEE container to enable combined applications such as the calling card example in Section 4.3.2.

The combined JAIN SLEE–JEE application has two parts: an enterprise application built from Java beans and a communications application built from SBB components. Given the scale of JAIN SLEE and JEE applications, it is likely that the JEE and JAIN SLEE part of the application run on different servers and hence different virtual machines.

Java offers several means for communications between objects in different containers and on different virtual machines, including remote method invocation (RMI) and web services. We won't go into these in detail here, but Chapter 5 explains the principle of web services.

In a similar way, SBBs can also connect to JEE components to create distributed communications applications with components in the network and on the terminal device. Figure 4.22 shows how JAIN SLEE can be combined with JEE and JEE.

The combination of JAIN SLEE and JEE allows for different types of applications, depending on where the primary point of control lies. If primary control lies in JAIN SLEE, the service provider has control of the service and

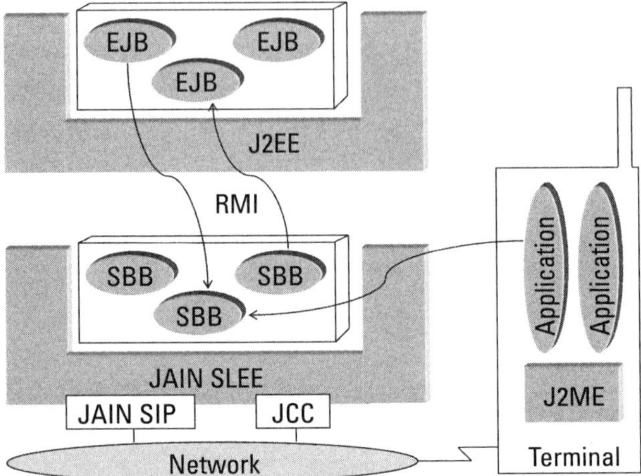

Figure 4.22   Combining JAIN SLEE, JEE, and JEE.

can provide carrier-grade performance. An example of a composed service where control lies with the service provider is a VPN service implemented in JAIN SLEE, where the service provider queries the enterprise user profiles from the enterprise's JEE application.

If primary control lies in JEE, the enterprise has control of the application and can ensure tighter control with enterprise back office applications. An example of a composed service where control lies with the enterprise is a click to call back service where the enterprise's JEE application decides from which office to call a customer who clicks a link on the corporate website to request a call from technical support staff. The JEE application instructs the JAIN SLEE application to set up that call from the designated office to the customer's phone.

4.6   Java IMS API

The JSIP API, SIP servlets, and JAIN SLEE differ in many ways, but they have one thing in common: they run in an application server. In this section we'll consider an API that facilitates the creation of applications on the device side. The Java IMS API provides an API for device-based applications that use the communication services of IMS.

The Java IMS API works with JEE, CLDC, and MIDP, the reduced Java run time environment for devices that have resource constraints (small processor and memory, small screen, and so on).

4.6.1 Capabilities

In the days of voice telephony, we took for granted that you could make a voice call between any two devices. The Internet, 3G mobile networks, and smart phones have made compatibility less trivial. Today's mobile devices have diverging characteristics and capabilities. They can have different peripherals (screen, camera, keyboard), can use different codecs, and can have different applications installed.

On the network side, different IMS networks may offer distinct services. We've seen in Chapter 3 that MMTEL and RCS try to align the basic communications services offered by IMS networks and improve interoperability between them, but different IMS networks may nevertheless support different services.

When building applications for IMS-enabled devices, compatibility between the device and network capabilities becomes critical. One of the most important features of the IMS API is to be able to specify, publish, and match the device and network capabilities so that the applications can actually connect to each other.

The IMS API recognizes two types of capabilities. Basic capabilities correspond to IMS-related media services such as multimedia telephony (MMTel), video sharing, and presence. Composed capabilities define features that are specific to certain applications (e.g., the possibility to locate the device or to exchange a specific type of XML document).

The 3GPP has defined two types of identifiers to denote standard and nonstandard IMS services:

- *IMS communication service identifiers* (ICSI) define IMS services that have been standardized by the 3GPP (e.g., `urn:urn-7:3gpp-service.ims.icsi.mmtel` denotes the MMTel service discussed in Section 3.5).
- *IMS application reference identifiers* (IARI) define IMS applications that haven't been standardized by the 3GPP as IMS services (e.g.,`rn:urn-7:3gpp-application.ims.iari.gsma-vs` denotes a GSM Association–defined video sharing application).

The IMS API specifications recommend application developers use standard IMS services wherever possible and register for new nonstandard applications with the IARI so that they are identifiable.

4.6.2 Registry

From the IMS API viewpoint, an application has two parts: the application-specific code and the IMS API–defined capabilities it uses. The latter is of critical importance to ensure compatibility between the device, the network, and other devices running the application. Each application therefore maintains a registry of properties that describe the IMS features it supports.

A property is a string containing a name and zero or more values. Many of the values are expressed as MIME types, which describe media types as strings of the form `<type>` or `<type>/<subtype>`. MIME types [11] are commonly used in e-mail headers to describe the content of the message and its attachments. Table 4.6 describes the basic properties used by the IMS API and provides examples for each.

Table 4.6
IMS API Properties

Property Name	Description	Example
Basic	Basic media transfer capability, using, for example, UDP over IP.	`application/x-doom`
Stream	Audio or video streaming capability.	`audio` `video` `video/h263`
Frame	Messaging capability.	`text/plain` `image/png`
Event	Event handling capability.	`presence`
QoS	Associates a flow specification (QoS parameters) with a media type. The flow specification is a seven-value vector that sets values for the average rate, buffer size, peak rate, delay, delay variance, max chunk size, and minimal policed size.	`{ Doom, application/x-doom, 1024 6000 2048 0 0 100 10 }`
CoreService	Associates the application with an IARI or ICSI.	`{ "myGame", "urn:IMSAPI:com. myCompany.myGame" }`

One of the important steps when installing an application on a device is to set the registry for the application, so the application can communicate its capabilities to its peers. The IMS API provides the static[6] class `Configuration` with the `setRegistry()` operation for this purpose. The application must call `setRegistry()` with three parameters: the application ID, the name of the class that implements the application, and the registry itself, all of which are strings. The following is an example of a registry definition and registration using the Java IMS API:

```
String[][] registry = new String[][] {
  { "Basic", "application/myGame" },
  { "Framed", "text/plain image/png" },
  { "Event", "Presence" },
  { "CoreService", "myGame", "urn:IMSAPI:com.
myCompany.iari.myGame", "", "" },
  { "Qos", "myGame", "application/myGame", "1024
6000 2048 0 0 100 10 ", "" },
};
Configuration.setRegistry("org.jsr281.myGame",
"MyGameImplementation", registry);
```

4.6.3　IMS API Structure

The IMS API provides applications with interfaces to

- Expose and negotiate capabilities.
- Connect to IMS, start sessions, and manage media connections. The IMS API provides high-level operations for managing IMS sessions and media connections without requiring the application to process individual SIP messages.
- Publish events and subscribe to events. The IMS API provides a publish-subscribe interface that follows the Observer pattern we saw in Section 4.1.5. Using the publish-subscribe interface, applications can receive network-related events from IMS but also application-specific events from other devices.

To achieve this, the IMS API provides three packages:

- `Javax.microedition.ims` defines the interfaces for checking the connection to IMS and creating the registry for an application.

- `Javax.microedition.ims.core` defines interfaces for querying capabilities, starting and stopping sessions, creating and parsing messages, and publishing and subscribing to events.
- `Javax.microedition.ims.core.media` defines the interfaces for managing media connections.

The IMS API specification JSR 281 [12], which is publicly available on the Internet, lists the classes in each package and their use. In the next section, we'll look at a few examples of how applications can use these classes.

4.6.4   Using the IMS API

To query the capabilities of another device, the application uses the `Service` interface of the `Javax.microedition.ims` package as a factory to create an instantiation of the `Capabilities` class, specifying the SIP address of the destination device as parameter of the `createCapabilities()` operation. The application then registers itself as an observer (listener) of the `Capabilities` instantiation and calls the `queryCapabilities()` operation on it:

```
Capabilities caps = service.
createCapabilities(null, "sip:president@
                    whitehouse.gov");
caps.setListener(this);
caps.queryCapabilities(false);
```

If the target device responds by sending its capabilities, the IMS API will call the `capabilityQueryDelivered()` operation on the application's `CapabilitiesListener` interface to notify it of the successful response, and the application can retrieve the capabilities from the `Capabilities` instance.

The next code snippet shows how an application can start a voice telephony session, assuming that the device has a voice over IP application (VoipApp) installed as core service:

```
myCoreService = (CoreService)Connector.open(
                  "imscore://com.someCompany.
                  VoipApp");
myCoreService.setListener(this);
mySession = myCoreService.createSession(null,
                  sip:president@whitehouse.gov);
mySession.setListener(this);
```

```
myMedia = (StreamMedia) mySession.createMedia(
        "StreamMedia",
            Media.DIRECTION_SEND_RECEIVE);
myMedia.setSource("capture://audio");
mySession.start();
```

In this code snippet, the call to `Connector.open()` instantiates a `CoreService` object that allows the application to expose its capabilities to the network and to initiate and receive IMS sessions. If the application wants to initiate a session, it must first create a `Session` object by calling the `createSession()` operation on the `CoreService` instantiation, providing the destination IMS or SIP address as parameter. The application registers itself as listener to the new `Session` object, so it can be notified of session-related events. At this point the application has created a Session object but has not yet defined the media parameters for the session.

To set the media parameters for the session, the application calls the `createMedia()` operation with the necessary parameters. In this example, it creates a `StreamMedia` object to enable bidirectional media streaming. The application sets the endpoint for the media stream by calling the `setSource()` operation on the `StreamMedia` object. In this case, the endpoint consists of the audio source and sink of the device—in other words, the microphone and speaker.

Now that the application has set the session and media parameters, it can start the session by calling `start()` on the `Session` object.

Applications can also respond to incoming session invitations by implementing the `sessionInvitationReceived` operation on their `CoreServiceListener` interface:

```
public void sessionInvitationReceived(CoreService
service, Session session) {
   /*The application can specify here what to do
with the incoming session.*/
};
```

In this section we've seen how the IMS API allows Java programmers to build device-side applications that can communicate over IMS, without having to manipulate individual SIP messages. The IMS API enables applications to check their compatibility with the IMS services used by the network and with the capabilities of peer devices before attempting to connect to them.

In this chapter we've dealt with only Java APIs to be used in Java applications. The next chapter will look at APIs from a different angle, as interfaces between distributed systems.

References

[1]   Alexander, C., S. Ishikawa, and M. Silverstein, *A Pattern Language: Towns, Buildings, Construction*, Center for Environmental Structure Series, New York: Oxford University Press, 1978.

[2]   Gamma, E., R. Helm, R. Johnson, and J. Vlissides, *Design Patterns: Elements of Reusable Object-Oriented Software*, Reading, MA: Addison Wesley, 1994.

[3]   Oracle, Connected Limited Device Configuration, version 1.1, JSR-000139 Expert Group, 2007.

[4]   Oracle, Mobile Information Device Profile for Java™ 2 Micro Edition, version 2.0, JSR 118 Expert Group, 2002.

[5]   Oracle, JSIP API Specification, version 1.2 (Final Release), JSR 32 Expert Group, 2006.

[6]   Oracle, SDP API Specification, JSR 141 Expert Group, version 1.0, 2004.

[7]   Kulkarni, M., and Y. Cosmadopoulos, SIP Servlet Specification, version 1.1, JSR 289 Expert Group, Oracle Corporation, 2008.

[8]   Oracle America, JavaMail™ API Design Specification, Final Release version 1.5, JSR-919 Expert Group, 2013.

[9]   Lim, S. B., and D. Ferry, JAIN SLEE 1.0 Specification, Final Release, Oracle Corporation, 2004.

[10]   Oracle, Java Call Control (JCC) Application Programming Interface (API), version 1.0b, 2002.

[11]   Freed, N., J. Klensin, and T. Hansen, Media Type Specifications and Registration Procedures, IETF RFC 6838, 2013.

[12]   Oracle, IMS Services API for Java Micro Edition, Version 1.0 (Final Release), JSR 281 Expert Group, 2008.

5

Application Programming Interfaces for Distributed Service Control

In Chapter 4 we saw how Java provides APIs that help the programmer create applications involving communications over IMS or legacy networks. Chapter 4 demonstrates how SIP servlets and JAIN SLEE SBBs can delegate all or part of the application processing to an enterprise domain by connecting to Java beans in a J2EE container. All software is programmed in Java, and the Java components have a relatively tight integration within a single deployment and business framework.

This chapter considers the use of APIs for the purpose of providing more loosely integrated applications that can span different software and business domains. In this case, the different components of the application can run on different servers, managed by distinct entities, and can be programmed in different languages.

Consider the example of a mobile virtual network operator (MVNO). A MVNO does not operate a network but does provide its own value-added services. To run its applications and services, the MVNO needs interfaces to components of the host network in order to manage registrations, control sessions, locate devices, and so on. Network APIs enable this deployment model. Network APIs allow third-party applications to access certain network

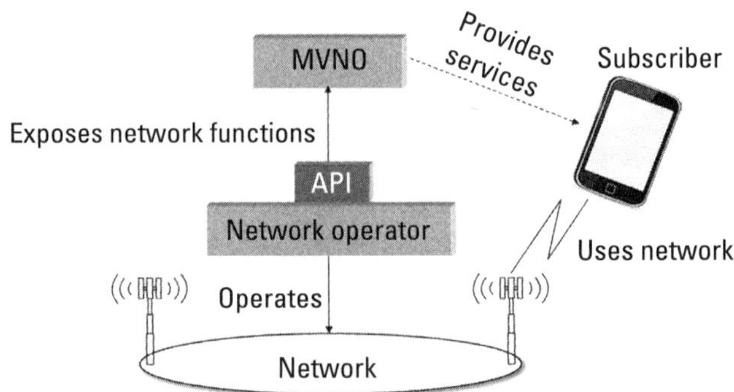

Figure 5.1   Third-party exploitation of network resources through APIs.

functionality, such as setting up calls, locating devices, and getting presence information for a subscriber, as illustrated in Figure 5.1.

Java API's are not the most ideal solution for this scenario, as the application software runs on different servers than the APIs, and the application software may be written in a different language than Java.

5.1   Distributed Object-oriented Systems

As we've seen in Section 4.1.1 and through the rest of Chapter 4, objects communicate by invoking operations on each other. An object calls an operation on another object by sending a request message specifying the parameters with which to execute the operation. When the operation terminates, it sends a respond message back with the result. There is no reason why request and response messages can't be sent between objects on different machines, as long as they have an Internet connection between them.

5.1.1   Communications Between Distributed Objects

A request message sent from an object on one machine to an object on another must at least

- Identify the object that the request message is intended for;
- Identify the operation to be called on that object;
- Specify values for the parameters of the operation to call.

This looks simple, but it's not as straightforward as it seems at first sight. First, the software on either machine must have a way to identify remote objects. Object references in Java and other languages are essentially pointers to memory. In the following line of Java code, the reference to `myObject` only has meaning on the virtual machine running this code:

```
myResult=myObject.myOperation(param1, param2);
```

A virtual machine on one machine has no knowledge of the memory on another machine and has no control over it.

Sending requests to remote objects on other machine requires a machine-independent way of identifying objects. The target machine must map this machine independent object identity to a local memory reference.

Another problem is how to pass the parameters because their types depend on the language they were defined in and the machine they run on. For example, Java defines a `short` integer as a 16-bit value, but other languages such as C++ allow `short` integers of more than 16 bits. If a C++ object on one machine were to call an operation on a Java object on another machine, passing a `short` integer of more than 16 bits as parameter, the Java object would have problems processing it.

Computer scientists have devised several solutions to address this problem. In the following sections, we'll briefly review them.

5.1.2 Remote Procedure Call

Programmers using the language C on Unix computers recognized the need for creating distributed applications before object-oriented programming became mainstream. They devised a solution called remote procedure call (RPC).[1] In RPC terminology, the program that calls the remote procedure is the client, and the remote program that runs the procedure is the server.

RPC introduces two components in a distributed application:

- A *stub* is a procedure on the client machine that has the same signature[2] as the remote procedure on the server but does not run the remote

---

[1] C is the default programming language on the Unix operating system (in fact, Unix itself is programmed in C). C is a procedural language where programs consist of a main body and procedures (sometimes also called *subroutines*) or functions—hence the term *remote procedure call*. A function is a simple kind of procedure that takes input-only parameters and produces one result value.

[2] A procedure's or function's signature consists of its name, parameter types, and result type (if it has one). For example, `int sum (int x, int y)` represents the signature of a function with name sum that takes two integer values as parameter and produces an integer value as result. Note that `int sum (int x, int y, int z)` has the same name but a different signature because

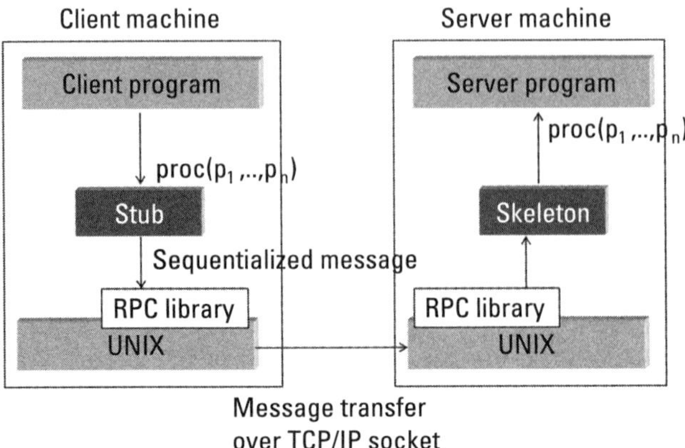

Figure 5.2   Remote procedure call concept.

procedure code. When called by the client program, the stub only creates a request message with the procedure's parameters provided by the client and sends that request to the server machine over a TCP/IP socket. The stub also receives a response message back from the server and passes the result back to the client.

- A *skeleton* is a procedure on the server machine that receives the request message over the TCP/IP socket and calls the requested procedure on the server program with the parameters defined in the request. When the procedure returns a result, the skeleton sends it back to the client in a response message.

The creation of a message that can be sent over the network from the parameters of a procedure is called *marshaling*, and the inverse operation (i.e., taking the parameters or result from a received message and making a procedure call or return with those values) is called *unmarshaling*.

One could say that the stub represents the server program on the client machine, and the skeleton represents the client program on the server machine. The stub makes the remote procedure look like a local procedure to the client program, which won't need to do marshaling or manage sockets. Likewise, the skeleton makes the client program look local to the server program. The

---

it has different parameters. In this case, we say that the function sum is overloaded because it has two distinct signatures. Most programming languages allow overloading because the compiler can see from the signature which function or procedure is referred to.

Table 5.1
Tasks of Stub and Skeleton (Simplified)

Stub	Skeleton
1. Receive procedure call from client program	1. Wait for incoming messages
2. Marshall procedure parameters into a request message	2. Receive request message on TCP/IP socket
3. Open TCP/IP socket to server	3. Unmarshal request message
4. Send request message	4. Call server procedure with parameter values extracted from request message
5. Wait for response message	5. Marshal result message into a response message
6. Receive response message	6. Send response message back to the client over TCP/IP socket
7. Unmarshal response message	
8. Return result of procedure call to client program	

server program just receives the procedure call as if its origin were local. Figure 5.2 illustrates the principle of RPC.

The stub and skeleton perform tedious but straightforward tasks that depend only on the signature of the procedure they represent. Table 5.1 gives an idea of what the stub and skeleton have to do. Note that step 4 on the stub relates to step 2 on the skeleton, and that step 6 on the skeleton relates to step 6 on the stub.

Since the tasks of the stub and skeleton only depend on the signature of the procedure, RPC can generate them automatically if provided with a specification of the signature. RPC provides the external data representation (XDR) and RPCGEN program for this purpose. XDR provides a stripped-down version of C that allows only for data type specifications and a brief section to define the signature of a procedure.

RPCGEN takes an XDR file and automatically generates the stub and skeleton. All the programmer has to do is compile the client program with the stub code, and the server program with the skeleton code. This enormously simplifies the task of writing distributed applications in C.

5.1.3 Java Remote Method Invocation

Java offers a mechanism very similar to RPC for creating distributed Java applications. The remote method invocation (RMI) API allows objects to call operations on remote objects that run on a different (virtual) machine.

The fact that Java is an object-oriented programming language gives it an advantage over C. Java has a built-in facility to define operation signatures: the interface (see Section 4.1.1). To create operations that can be called remotely, the programmer defines them in an interface that inherits from Java's `Remote` interface to tell the compiler that it has to generate the corresponding stub and skeleton. The following example line tells the Java compiler that classes implementing the `MyService` interface provide operations that can be called remotely:

```
public interface MyService extends Remote { … }
```

A class that implements the `MyService` interface will have to inherit from the `UnicastRemoteObject` class, so that the Java compiler will generate a stub and skeleton for it.

```
public class MyServiceImplementation
        extends UnicastRemoteObject
        implements MyService { … }
```

The programmer does not have to worry about stubs and skeletons: Java generates them automatically.

The fact that Java is object-oriented also introduces a few technical issues that did not exist in C:

- *Finding and referencing remote objects.* As in RPC, a client object references a remote object through a stub, as if it were a local object. This is convenient for the programmer, but RMI has to take several steps to make this possible. The stub must have a remote reference that identifies the remote object on the server machine. The skeleton must be able to translate this remote reference to a local reference on the server machine. RMI does all this behind the scenes. As an additional facility, RMI lets a programmer assign names (strings) to objects and search objects by name at run time, using the naming service.
- *Garbage collection.* Java's garbage collector keeps track of object references and frees up memory occupied by objects that are no longer referenced and used by other objects. The garbage collector avoids memory filling up with junk as an application executes. RMI needs the Java garbage collector to keep track of remote object references, too.
- *Remote exceptions.* In object-oriented programming, exceptions provide a way to catch and handle errors before they can lead to critical

failure. In RMI, exceptions can also occur in the remote object, and the client object will have to know about them.

RMI provides the Java programmer with the necessary tools to build distributed Java programs. The Java compiler takes care of the generation of stubs and skeletons, and the run time environment takes care of the administration of object references, garbage collection, and remote exceptions. But what if the client and server have to be programmed in different languages? Imagine for example a system where a Java client needs to call an operation on a C++ server. The next section addresses this problem.

5.1.4　Common Object Request Broker Architecture

Java RMI enables distributed Java applications with components running on different virtual machines. An additional problem arises when not all of the components of a system are implemented in Java. Many large-scale software projects involve legacy systems programmed in older and sometimes non–object-oriented languages that are too large, too complex, or too costly to reimplement in Java.

Take the example of banking computer systems. The account management software of many banks still runs on mainframe computers and uses legacy languages such as COBOL. This mainframe software is complex, mission critical, and very costly to change. The software for the desktop clients in bank offices and teller machines is more recent and programmed in C++, C#, Java, or other object-oriented languages.

Staying closer to the world of communications, many operators (especially incumbent operators) still rely on legacy billing systems from the 1980s and 1990s that are too costly to replace. Yet, as we've seen throughout this book, their network has evolved to support state-of-the-art IMS-based communications. The integration of newer network systems with older billing systems programmed in legacy programming languages can create a real challenge. RPC and RMI do not provide a straightforward solution for this problem because both are bound to a particular programming language.

The object management group (OMG) specified the common object request broker architecture (CORBA) in 1991, enabling programmers to create distributed object-oriented applications across languages and computer platforms. The OMG is an industry consortium setting standards for object-oriented design and programming. Observe that CORBA in fact came before Java and RMI, which date from the second half of the 1990s.

CORBA [1] works with stubs and skeletons, and builds on the same principles as RPC and RMI. However, CORBA tackles a couple of additional challenges that do not exist in distributed applications programmed in a single language.

The first question is how to define an interface of a remote object, given that CORBA has to work with multiple programming languages that declare interfaces differently. What's more, the OMG specified CORBA to work also with legacy programming languages such as COBOL and C, which don't have objects and interfaces at all!

5.1.5   Interface Definition Language

To overcome this problem, CORBA has its own language for specifying interfaces, called the interface definition language (IDL). IDL allows the programmer to use data types that are common in most programming languages such as integers, floating point numbers, strings, and arrays, but it does not support the more exotic data types found in some languages. The interface of the example in Section 4.1.2 would look like this in IDL:

```
interface MultiPartyCall: VoiceCall {
  void addParty(in E164Num p);
  void dropParty(in E164Num p);
}
```

Although this example may look deceivingly like its Java counterpart in Section 4.1.2, there are some differences. Most importantly, the interface definition explicitly marks the parameters of operations as input (`in`), output (`out`), or mixed (`inout`) parameters. If this looks strange, remember that CORBA has to work with languages that allow procedures to modify their parameters, such Fortran and Pascal. Another difference is that CORBA uses the semicolon instead of the keyword `extends` to denote inheritance. CORBA and Java interface definitions differ in other ways, too, but we won't go into this in further detail.

To generate stubs and skeletons for the languages it supports, CORBA uses a similar approach as RPCGEN, which we saw in Section 5.1.2. CORBA refers to these generators as IDLtoJava, IDLtoC++, IDLtoCOBOL, and so on. Figure 5.3 shows how to integrate a client written in Java with a server written in COBOL.

After specifying the interface of the remote object in IDL, the programmer runs IDLtoJava to generate the Java stub and skeleton, and IDLtoCOBOL

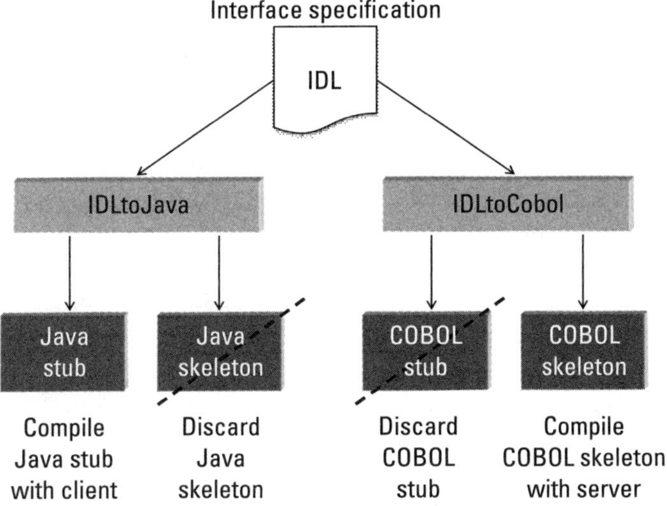

Figure 5.3   Stub and skeleton generation with CORBA IDLtoX.

to generate the COBOL stub and skeleton. As the client is programmed in Java, the programmer compiles the Java stub with the Java client while discarding the Java skeleton. On the COBOL side, the programmer compiles the COBOL skeleton with the COBOL server application and discards the COBOL stub.

5.1.6   Remote Object References

Another issue that CORBA addresses is how to reference remote objects in a language-independent way. CORBA imposes its own object references, allowing a client to access even procedures in a non–object-oriented language as if they were objects. To do this, CORBA creates a fictitious remote object reference for the procedure. A CORBA object reference is an opaque data structure for internal CORBA use. An application can use it to access a remote object, but it cannot interpret or process the remote object reference.

Like RMI, CORBA has a naming service that allows servers to associate remote objects with a string and lets clients to look for them by name. CORBA also provides additional features such as remote object persistence[3]

---

[3] A persistent object is an object that automatically saves its state in a database. This allows objects to survive their run time environment. Persistence is also used by applications to deactivate the object when it's not being used and reload it when it is needed again.

and transactional processing.[4] CORBA has its own protocol for the communication between distributed objects, the Internet interoperation protocol (IIOP).

CORBA facilitates distributed object-oriented programming, integrating components across languages and platforms, but it also has drawbacks. CORBA needs all distributed servers to run the object request broker (ORB) software and to exchange information about their interfaces and classes. IIOP uses specific TCP ports and a binary message format that has no compatibility with other protocols or systems. CORBA tightly manages its remote object references, which are meaningless outside a particular CORBA deployment.

These are not problems when building a distributed system within a single administrative domain, but for more loosely coupled systems these integration requirements are too restrictive. By using plain text request and response messages and standard protocols such as HTTP, web services (WSs) provide a more open framework to build loosely integrated distributed systems. In the next section, we'll take a more detailed look at WSs.

5.2   Web Services

The idea behind WSs is to make distributed applications communicate in a similar way that we surf the web for information. One could liken RPC, RMI, and CORBA to the situation where you know the exact URL of the web server that holds the information you need and you know exactly how to find the information on that server. In reality, of course, we tend to use the web in a different way. First, we use a search engine to find what we're looking for with a few keywords. Then we try the links the search engine presents us with to see if they hold the information or service we need.

WSs work in a similar way, by letting applications search for the remote service they need and then connecting to it. What's more, WSs use the same protocol (HTTP) and the same plain text message format (XML) as the WWW. WSs strongly rely on XML, so we'll first review this language.

---

[4] A transaction is a set of operations that have to be must be executed together (i.e., either all or none of the operations must complete successfully). Consider the example of purchasing a downloadable item from a website. In this case the transaction consists of two operations: pay for the item and download it. If payment is not successful, the user should not be able to download the item. On the other hand, if the product cannot be downloaded (e.g., due to server error), the user should not be charged for it. A transaction server monitors that all operations in a transaction successfully terminate; if one or more operations don't end successfully, the server automatically rolls back (undoes) the operations that were already completed.

5.2.1 XML

The extensible markup language (XML) is a language for structuring information in plain text format. XML [2] uses tags delimited by the < and > characters to structure information in a document. An element in an XML document consists of an opening and closing tag and possible text between it.

The following is an XML document meant to describe information about a book:

```
<?xml version="1.0"?>
<book>
   <author>Johan Zuidweg</author>
   <title>
   Next Generation Intelligent Networks
   </title>
   <isbn> 9781580532631</isbn>
   <publisher>Artech House</publisher>
   <year>2002</year>
   <hardcover></hardcover>
</book>
```

This snippet of XML shows that XML documents are human readable and can be self-explanatory if the tags are well chosen. The first line of the document contains the XML declaration. It's optional, but almost all XML documents include it. The example shows the elements delimited by tags may contain any plain text and may also have no content. In the example here, `<hardcover></hardcover>` represents an empty element explaining that the book is published with a hard cover. This empty element can be abbreviated as `<hardcover />`.

XML is extensible because the language allows the use of any tags. In this respect, XML differs from other markup languages such as HTML, which has a fixed set of allowed tags. XML has a very simple syntax with only a few rules:

- An XML document must have a single root element. In the previous example, the `<book>` tag denotes the root element. The root element encapsulates all other elements.
- All opening tags must have a matching closing tag.
- Tags must nest properly. For example, `<tag1><tag2></tag2></tag1>` is correct but is `<tag1><tag2></tag1></tag2>` is not.
- XML is case sensitive. For example, `<title>`, `<Title>`, and `<TITLE>` all denote different elements.

XML allows tags to have attributes to qualify the element they represent. For instance, the previous example could use the tag `<ISBN type="13">` to show that the element it holds a 13-digit ISBN number. Attribute values must be quoted, as this example shows.

Many applications need a more specific definition of the syntax of its data than what XML provides. Take, for example, the XML-based language CPL we saw in Section 3.4.7. CPL uses specific tags and attributes, and checking whether a document contains valid CPL requires rules beyond the basic XML syntax. Document type definitions (DTDs) and XML schemas provide alternative ways to impose additional syntax on XML documents. Although DTDs are somewhat easier to use than XML schemas, they have less expressive power and they have the added disadvantage that their text format differs from XML. XML schemas provide syntax for XML documents in XML. As WSs use XML schemas, we'll study them in more detail.

5.2.2   XML Schemas

An XML schema [3] defines which elements may appear in an XML documents, how to nest and order them, and which attributes and content they may have. XML schemas are themselves defined in XML. An XML document conforms to an XML schema if it strictly follows the syntax described by the schema.

Although XML schemas have a straightforward definition, reading, and writing, a correct XML schema is not that easy and requires meticulous description of the elements, their types, and order. XML schemas can become particularly complicated if elements have composed content types.

We'll not go into the details of XML schemas here because W3C provides a complete definition of XML schemas online,[5] and you can find countless tutorials on the web.

To get a taste for what a schema looks like, consider the following example of an XML schema defining how to construct information about a book. The XML example in Section 5.2.1 conforms to this schema:

```
<?xml version="1.0"?>
<xs:schema xmlns:xs="http://www.w3.org/2001/
XMLSchema">
<xs:simpleType name="ISBN13">
```

---

[5] In keeping with their mission, the World Wide Web Consortium (W3C) publishes their specifications on their portal, w3.org, in HTML format.

```
            <xs:restriction base="xs:stoken">
                <xs:pattern value=" [0-9]{13}"/>
            </xs:restriction>
        </xs:simpleType>
        <xs:element name="book">
          <xs:complexType>
            <xs:sequence>
                <xs:element name="author" type="string" />
                <xs:element name="title" type="string" />
                <xs:element name="isbn" type="ISBN13" />
                <xs:element name="publisher" type="string" />
                <xs:element name="year" type="gYear" />
                <xs:element name="hardcover">
                    <xs:complexType />
                </xs:element>
            </xs:sequence>
          </xs:complexType>
        </xs:element>
    </xs:schema>
```

This example illustrates how the `<xs:element name="…",
type="…">` tag defines an element with its name and content type, and
how the `<xs:sequence>` tag specifies a sequence of elements. The XML
schema requires any element that has nested content to be defined as a
`<xs:complexType>`, so `<xs:sequence>` must usually be defined as part
of a `<xs:complexType>`. Note how the schema defines `<hardcover>` as an
empty element by declaring it as an element of a complex type without content.

This example also shows that elements can contain strings as well as
XML predefined types such as `gYear`, which specifies a year as four digits
(e.g., 2015). XML schemas allow the programmer to define his or her own
data types from predefined types. The XML schema in the earlier example
uses a regular expression[6] to define the type `ISBN13` as a series of 13 digits.

---

[6] A regular expression is a construct from formal language theory that describes what is called
a *regular grammar*. A regular expression constructs a set of strings (finite or infinite) from
a finite set of literal characters (e.g., {"a," "b," "c"}), and three operations: concatenation (usu-
ally represented by ·), union (usually represented by |), and Kleene star (0 or more repetitions
of an element, usually represented by *). For example, the regular expression a·[b|c]*·a
generates the strings aa, aba, aca, abba, acca, abca, acba, abbba, … (an infinite set).

A regular expression can be represented by a finite state machine—that is, a finite graph consisting

In this example, you may have noticed that all tags have an `xs:` prefix. This prefix denotes an XML name space. A name space defines a unique syntactic prefix that avoids ambiguity between elements. Namespaces become a necessity when different XML documents use elements with the same name but with different meaning. For example, one XML document may use the element `<title>` to denote the title of a book, while another XML document may use `<title>` as the title of the author ("Professor," "Dr.," and so on). If we now build a library application that has to combine both XML documents, the meaning of the element `<title>` becomes ambiguous. Prefixing the element name with a name space avoids this ambiguity (e.g., we could use `<a:title>` to denote author titles and `<b:title>` to denote book titles). An XML schema can define the name space with a uniform resource identifier (URI)[7] that makes it unique (e.g., `xmls:a=www.johanzuidweg.com/authors`).

The root element of the schema `<xs:schema xmlns:xs="http://www.w3.org/2001/XMLSchema">` defines the name space `xs` as `http://www.w3.org/2001/XMLSchema`. Note that the URI simply represents a string that distinguishes `xs` in a globally unique way. The schema does not rely on information in the URI, which does not even have to correspond a real HTTP server. XML name spaces are a purely syntactical construction.

An XML validator is software that checks whether an XML document is well formed and whether it conforms to a given XML schema. An XML validator takes the XML document and an XML schema as input and produces a Boolean result (pass or fail), as illustrated in Figure 5.4.

Figure 5.4 also shows that XML schema validators can check whether an XML schema conforms to the W3C schema definition [4, 5]. XML validators and XML schema validators usually also give warnings and other information additional output to help the programmer improve the structure of the XML document. Figure 5.4 does not show this additional output.

5.2.3 Simple Object Access Protocol

Perhaps the most important principle behind WSs is that they take advantage of existing textual request-reply protocols to make requests to remote objects and receive results back from them. WSs most commonly use HTTP as its

---

of states (drawn as nodes) and transitions between them (drawn as arrows labeled with a literal character).

[7] A uniform resource identifier (URI) is a string that identifies a web resource and consists of two parts: the URI scheme and a scheme-specific string, separated by a colon. Well known URI schemes are `HTTP:`, `HTTPS:`, `SIP:`, `MAILTO:`, and `FTP:`.

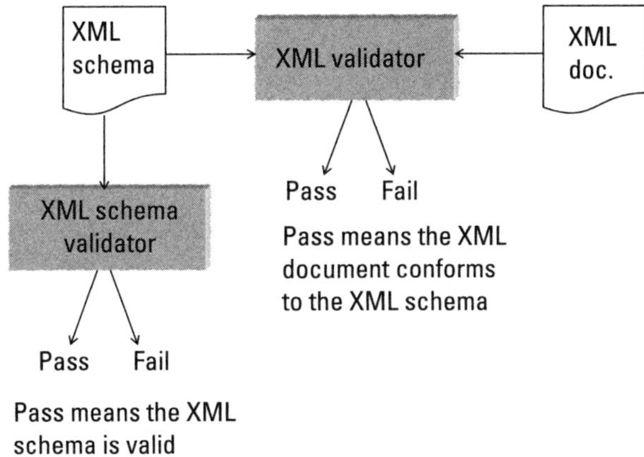

Figure 5.4   XML validator and XML schema validator.

request-response protocol, but can also use SMTP, FTP, or any standard protocol capable of sending text messages between computers.

Apart from its simplicity, HTTP brings important advantages. Almost all computers, tablets, and smart phones in the world support HTTP because all web browsers have a built in HTTP client. And HTTP uses a TCP port (number 80) accepted by almost any network, even networks with tight security, because web browsing is a basic necessity.

WSs use XML to carry the request and response data in HTTP messages. WSs refer to this XML payload as the single object access protocol (SOAP), although it's not really a protocol in its own right but rather a message format. Figure 5.5 shows the structure of a SOAP message, assuming HTTP as the host protocol.

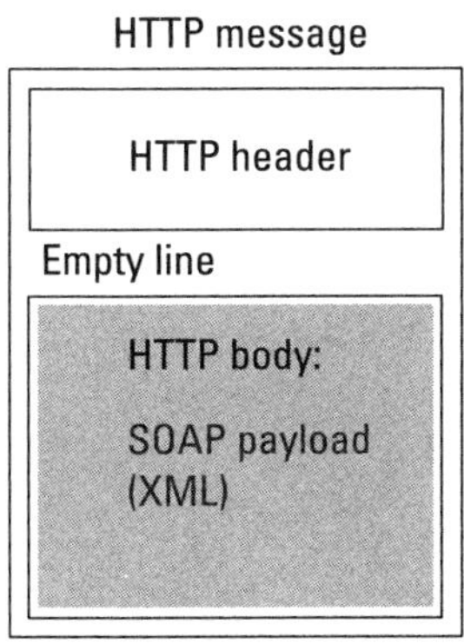

Figure 5.5   Structure of a SOAP message.

HTTP has two types of messages: requests and responses. A HTTP message consists of a mandatory header and an optional body. The first line of a HTTP request header identifies the request method. HTTP has eight types of request methods, among which the most frequently used are GET and POST. A GET request asks the server to return the (HTML) content at the specified path, while a POST message is used to send the content in the HTTP body to the server. The first line of a HTTP response header contains a three-digit response code (e.g., 200 OK, which indicates that the request was received and processed successfully).

An empty line separates the header from the body, which may or may not have plain text content. SOAP uses the HTTP request or reply body to transport the WS request or reply.

A SOAP message is an XML document with root element <Envelope>. The <Envelope> element contains an optional <Header> and a mandatory <Body> element.

The header can contain meta-information telling the server how to process the message (e.g., requesting execution at a specific date and time). In practice, WS requests rarely use the header.

The SOAP body carries the actual request or response. In the case of a request, the body describes the operation to be called an XML element and the parameters of the operation children elements. In the case of a response, the body of the SOAP message contains the result of the operation. Although SOAP is XML and HTTP is a plain text protocol, SOAP messages can carry nontext parameters by MIME encoding them (see Section 3.1.2).

The following is an example of a SOAP request (note that this request resembles a call to the connect operation of the voiceCall interface of Section 4.1.1):

```
POST /callServer/voiceCall HTTP/1.1
Host: 209.110.197.2
Content-Type: text/xml
Content-Length: 277
<?xml version="1.0"?>
<soap:Envelope
 xmlns:soap="http://www.w3.org/2001/12/
soap-envelope">
  <soap:Body xmlns:c="http://www.johanzuidweg.com/
call">
    <c:connectRequest>
      <c:caller>696858585</c:caller>
```

```
      <c:callee>696904123</c:callee>
    </c:connectRequest>
  <soap:Body>
<soap:Envelope>
```

The first four lines contain the HTTP header. The POST line identifies the path on the server to the remote object that has to process this request—in this case, /callServer/voiceCall. The Host: line identifies the client of the request—in this case, the machine with IP address 209.110.197.2. The third line shows that SOAP messages are of type text/xml, and the fourth line gives the content length in bytes.

The remainder of the message beginning with the line <?xml version="1.0"?> describes the SOAP request. The <Envelope> root element declares two name spaces. The soap: prefix refers to the W3 defined name space for SOAP elements, while the c: prefix refers to a service-specific name space. Although XML and SOAP do not require it, the use of name spaces is common and highly recommended.

The SOAP message in this example contains a body but no header, which is also common. The <soap:Body> element contains the <c:connectRequest> element identifying the operation to be called on the remote object with <c:caller> parameter value 696858585 and <c:callee> parameter value 696904123.

SOAP response messages have a similar structure, as illustrated by the following example:

```
HTTP/1.1 200 OK
Content-Type: text/xml
Content-Length: 359
<?xml version="1.0"?>
<soap:Envelope
 xmlns:soap="http://www.w3.org/2001/12/
soap-envelope">
  <soap:Body xmlns:c="http://www.johanzuidweg.com/
call">
    <c:connectResponse>
      <c:connection>
              http://www.johanzuidweg.com/
callService/connection/rbcC397ytRv2y543
      </c:connection>
```

```
      </c:connectResponse>
   <soap:Body>
 <soap:Envelope>
```

Note that this example deviates from the interface example in Section 4.1.1 because there the operation `connect()` had a void response. To make this example a little more interesting, we assume that the `<c:connect>` message returns a reference to a `<c:connection>` object specified as a web service by an URL.

This section has shown how SOAP provides a straightforward way to encode requests to and responses from remote objects in XML, using HTTP as a host protocol to transport the messages. As in any other object-oriented system, a WS client needs to program to the interface of the remote object. In other words, WS need a way to specify interfaces of remote objects. WS provides the web services definition language (WSDL) for this purpose. Most people in the trade pronounce the somewhat awkward acronym WSDL as "wisdle."

5.2.4 Web Services Definition Language

A WSDL document describes the operations a client can call on a web service. In other words, it defines the interface of the remote objects that implement the web service. The format of WSDL documents is XML, and as such it has one root element: `<definitions>`. Under the root element, a WSDL document contains four types of elements:

- `<types>`: this element contains an XML schema to define data types used by the web service. For example, this element can describe how a client must structure a customer record consisting of name, address, e-mail, and phone number in XML.
- `<message>`: this element defines the request and response messages and their parameters.
- `<portType>`: this element specifies the message flow. The normal message flow for calling operations on remote objects is request-response, but WSs do allow for other message flows such as request only (no response) or response only (no request).
- `<binding>`: this element defines which protocol and message encoding to use for a port type. The default used in most cases is HTTP, but WSDL allows the use of other protocols such as SMTP or FTP, too.

The best way to understand the structure of a WSDL document is by looking at an example. The following WSDL document specifies the web service that allows clients to send the SOAP request and receive the SOAP response of Section 5.2.3. To avoid too much clutter for human reading, this example does not show name spaces:

```
<?xml version="1.0"?>
<definitions>
 <types>
  <schema>
   <simpleType name="e164Num">
     <restriction base="xs:stoken">
         <pattern value="[0-9]{9}"/>
     </restriction>
   </simpleType>
  </schema>
 </types >
 <message name="connectRequest">
   <part name="caller" type="e164Num"/>
   <part name="callee" type="e164Num"/>
 </message>
 <message name="connectResponse">
   <part name="connection" type="string"/>
 </message>
 <portType name="voiceCall">
  <operation name="connect">
    <input message="connectRequest"/>
    <output message="connectResponse"/>
  </operation>
 </portType>
</definitions>
```

The `<definitions>` element contains an XML schema that defines `<e164Num>` elements as series of nine digits.[8] The two `<message>` elements

---

[8] This is, of course, not the correct definition of an E.164 number as specified by the ITU-T. The ITU-T specification allows E.164 numbers to have variable length up to 15 digits and to begin with the "+" sign. Here, we've used a simplistic definition in order not to make the XML schema too long and complex.

define the format of the `<connectRequest>` and `<connectResponse>` messages. The `<connectRequest>` message takes two parameters of type `e164Num`, and the `<connectResponse>` takes one parameter of type `string`.

The `<portType>` element then defines the connect operation as a request-response interaction where `<connectRequest>` is the input message (request) and `<connectResponse>` is the output message (response). This example omits the `<binding>` element for simplicity. If used, `<binding>` tends to be the most complex WSDL element, but in the vast majority of cases WSs will use a default HTTP binding.

This example shows that WSDL files are tedious to read (even more so had we included the `<binding>` element and all the name spaces) but rather straightforward. Remember that although both WSDL and SOAP have plain text formats, their principle target is machine-to-machine communication, not human reading and enjoyment.

5.2.5 Representational State Transfer

Compared with RMI and CORBA, SOAP and WSDL provide a language- and platform-independent way to make remote objects communicate, taking advantage of standard Internet protocols and formats such as HTTP and XML. But on the downside, WSs don't facilitate the generation of stubs and skeletons from an interface definition, as RMI and CORBA do. The programmer has the responsibility to provide the software that parses SOAP requests and assembles SOAP responses according to a WSDL specification on the server side, and vice versa on the client side. Many Java, C++, and C# libraries exist that facilitate programming with WSDL and SOAP, but they have a significant learning curve.

Reducing things to their essence, one can classify the interactions between objects as one of the following basic actions:

- Create a new object.
- Delete an object.
- Query the state of an object (i.e., get one or more of its attribute values).
- Modify the state of an object (i.e., change its attribute values by calling one of its operations).

This made computer scientists realize they could apply R. Fielding's representational state transfer (REST) principles [6] to WSs. REST describes a standard way to access web resources, with the intent of simplifying web interactions and reducing the risk of unpredictable side effects. The application of REST principles does not give rise to new WS standards; it merely describes a structured and simpler way to publish and use WSs.

Although REST has academic origins, the specification of the HTTP 1.1 protocol builds on its principles.

In Section 5.2.3 we saw that WSs commonly use the HTTP POST request to send SOAP requests to a server, regardless of the content of the SOAP request. If a SOAP request intends to get some value from a remote object, REST principles consider it more appropriate to use the HTTP GET request, so that the semantics of the HTTP method matches the intent of the request. Likewise, if a SOAP request has the purpose to delete a remote object, REST considers the use of the HTTP DELETE request as more appropriate.

REST treats any object that can be accessed through HTTP as a resource[9] and defines four types of operations on web resources. These four types of operations coincide with the HTTP methods GET, POST, PUT, and DELETE, as shown in Table 5.2.

The GET and DELETE methods need to identify only the directory or resource to be retrieved. The POST and PUT requests, on the other hand, also need to specify the attributes of the resource to be created or modified. There are two ways to do send resource attributes with a HTTP request:

- *In a query string:*[10] this is the simplest way of passing attributes with a HTTP request and avoids having to send data in the body of the HTTP message. However, query strings are only apt for simple attributes such as integers and strings, and become cumbersome when the attributes have complex data structures.
- *In the HTTP message body:* this is also how SOAP makes WS requests. REST allows HTTP messages to carry SOAP or JSON payloads.

---

[9] This is another fine example of terminology clutter. The distributed object-oriented programmer's community usually refers to distributed objects as *remote objects*. The web service community tends to use the more general term *service*, and the REST community refers to *resources*. Although it's true that object-oriented computing, web services and REST have distinct scope and purpose, an object is still an object and a rose by any other name smells just as sweet.

[10] A query string provides a standard mechanism to parameterize an URL. The general use of query strings is in URLs of the form `http://server/path/resource?query_string`, where the `query_string` has the format `parameter1=value1¶meter2=value2&...` (all in plain text). Query strings are commonly used to pass parameters to Java servlets, CGI, or PhP programs.

Table 5.2
The Use of HTTP Request Methods in RESTful WSs

HTTP Request	Effect on Resource
GET	Return the resource identified in the request URL, in XML or Javascript object notation (JSON)[1] format.
PUT	Replace the attributes of the resource as specified in the request. If no resource exists as specified in the request URL, create a new remote object with the specified attributes.
POST	Create a new resource in the specified directory path and return its URL.
DELETE	Delete the resource as identified in the request URL.

[1] The Javascript object notation (JSON) provides an extremely simple format for representing objects as `"name":value` pairs, using curly brackets `{}` for enclosing and nesting. An example of a JSON string that describes the connection object created in the example of Section 5.2.3 is `{"caller":696858585,"callee":696904123}`. Compared with XML, JSON provides for more compact textual structures, but XML has richer expressive power. Remember that Javascript and JSON have nothing to do with the programming language Java (see the footnote on Javascript in Chapter 4!).

JSON provides a simpler and more compact format than SOAP but has less expressive power. SOAP and JSON payloads are more appropriate for HTTP messages that carry complex attribute values.

Using HTTP semantics explicitly to describe the action on the remote object, the WS interface only has to treat the remote object as a REST resource identified by an URL. The general format to reference a remote object as a REST resource is:

```
http://server/path/resource
```

where `server` identifies the server URL (e.g., `www.johanzuidweg.com`), `path` identifies the directory on the server where the server can be found (e.g., `/voiceCall/connections`), and `resource` names the remote object. An example of a resource URL is `http://www.johanzuidweg.com/callServer/connections/650jHuy43rc9Vbuo`, where the string `650jHuy43rc9Vbuo` identifies a particular connection object.

REST requires HTTP interactions to have no state. In other words, a client should not rely on the server to keep the state of a series of HTTP transactions. Every transaction must be fully independent from the others.

5.2.6   A REST Example

If all this looks a bit abstract and puzzling, maybe it's a good time to look at an example. Let's go back for a moment to the example of the voice call remote object with the operation `connect` that takes two parameters of type `e164Num`, of which we've seen the SOAP request and response examples and the WSDL definition in Sections 5.2.3 and 5.2.4 respectively. With REST, we can access `voiceCall` objects through the URL `http://www.johanzuidweg.com/callService/voiceCalls/` and `connection` objects through the URL `http://www.johanzuidweg.com/callService/connections/`.

In the example of Section 5.2.3, calling the `connect` operation creates a `connection` object. According to Table 5.2, the `POST` method has the semantics of creating a new object, so with REST we can make the `connect` request by sending a HTTP `POST` request to the URL:

```
http://www.johanzuidweg.com/callServer/connections
?caller=696858585&callee=696904123
```

This HTTP message does not need to carry any information in its body because the `POST` method already tells the server to create a new connection object, and the query string[10] of the HTTP request provides its parameters.

With REST, retrieving the list of all active connections becomes as simple as sending a HTTP `GET` request to the URL `http://www.johanzuidweg.com/callServer/connections/`.

Assuming that the connection created earlier has the identity `650jHuy43rc9Vbuo`, a client can change the connection's destination address (which effectively forwards the call to a new destination) by sending a HTTP `PUT` request to the URL `http://www.johanzuidweg.com/callServer/connections/650jHuy43rc9Vbuo?callee=692323591`.

And REST allows for deleting this connection by sending a HTTP `DELETE` message to `http://www.johanzuidweg.com/callServer/connections/650jHuy43rc9Vbuo`.

This example shows how REST can eliminate the need for SOAP or JSON payloads in HTTP messages, by explicitly using HTTP semantics and passing operation parameters as query strings.

So does REST have nothing but advantages over SOAP and WSDL? It depends. Taking a simplistic view, one could say that REST puts web resources at the center, whereas SOAP puts more emphasis on describing operations. REST provides a particularly efficient way of getting and setting object values or creating and deleting objects. But if the remote object has operations with complex parameters, it becomes cumbersome to encode them as query strings. REST does admit the use of XML or JSON in the HTTP body to encode complex parameters, but this also weakens its advantage over SOAP.

Unlike SOAP and WSDL, REST-based WSs have no formal specification. There is no standard that defines how the server tells the client what parameters it can accept with HTTP requests. Nor is there a standard format in which to return a resource as the result of a GET request from a client. REST allows a server to return a resource description in SOAP or JSON, but does not specify how. In practice, service providers specify their RESTful WSs with WSDL, JSON, or, very often, plain text.

Now that we've seen the mechanics of APIs, we can start looking at the APIs the industry has defined to build communications applications.

5.3   The Emergence of Network APIs

As explained in the introduction of this chapter, APIs have the purpose of providing third-party access to data or network control features owned and managed by the network operator service provider. The service provider decides which control features to delegate to third parties by exposing part of the network functionality as an API. Figure 5.6 shows how APIs can delegate control to third-party applications in both circuit switched (IN, CAMEL) and packet switched (IMS) communications networks.

In the left-hand side of Figure 5.6, the SCF does not process the call but instead provides an API through which a third-party application can control the call. If the API provides an abstract view of the call, the third-party application does not need to be aware of the underlying protocols or the network.

The right-hand side of Figure 5.6 shows a similar deployment in IMS, where an AS provides an API to third-party applications to control multimedia sessions. In this case, too, the third-party applications are programmed against an interface and do not need to process SIP messages.

Service providers can also provide APIs for other network resources such as short messaging centers (to send and receive short messages), multimedia messaging centers (to send and receive multimedia messages), mobile location servers (to locate a subscriber), or HSS (to query and manage subscriber data).

Several standards organizations including 3GPP, GSMA, and the Open Mobile Alliance (OMA) have defined network APIs. In this chapter, we'll study these APIs in more detail, but before doing so it may be useful to understand their history and how they relate.

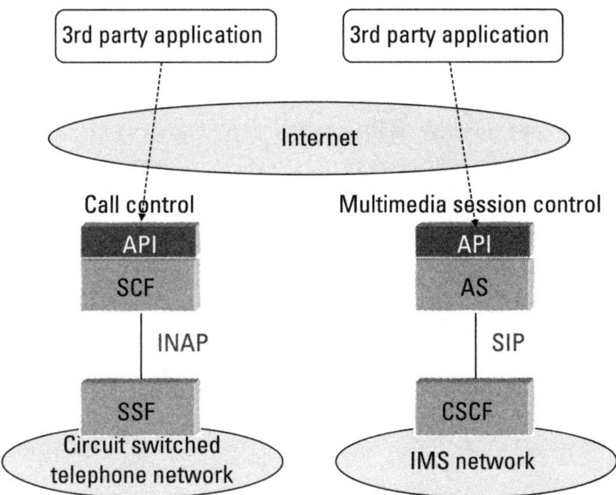

Figure 5.6 Delegating call control or multimedia session control through APIs.

5.3.1 The Parlay Group

One of the first initiatives aimed at defining network APIs stems from the late 1990s. At that time, the circuit switched telephone network still represented the mainstream in telecommunications, the WWW was an emerging infrastructure, and IMS had not yet been defined. But it was also a time of worldwide privatization of the telecommunications market and fast technological evolution in the Internet domain.

The R&D departments of network operators and equipment vendors had already begun experimenting with distributed object-oriented programming applied to telecommunications services. In the early 1990s, key players in the telecommunications ecosystem created the large scale Telecommunications Information Networking Architecture Consortium (TINA-C) [7] to design a disruptive telecommunications architecture based on distributed object-oriented programming technology (in particular, CORBA; see Section 5.1.4). Although no operator eventually deployed TINA commercially, TINA-C certainly did represent a trend toward more open telecommunications architectures based on state-of-the-art information technology.

In a move typical for the turbulent telecommunications scene of the time, the UK Office of Communications (Ofcom)[11] announced that incum-

---

[11] The UK Office of Communications regulates the communications industry, including broadcast radio and TV and the postal service. At the end of the 1990s, it went by its previous name Oftel.

bent operator British Telecom (BT) would have to allow third-party service providers access to its telephone exchanges so as to facilitate competition. BT obviously struggled with this impending regulation, both from a technological and a business point of view. BT responded by creating the Parlay[12] Group in 1998 with a number of associates from industry (initially Microsoft, Nortel, Siemens, and Ulticom). The Parlay group published its first API specifications the same year. The Parlay API [8] provides third-party access to the network's call control features and effectively opens the interface between the IN SSF and SCF (see Section 2.5.1). Parlay also specified a framework to authenticate and authorize third parties to use all or part of the API, so as to protect the integrity of their network and to be able to charge the involved parties for the network resources used. Parlay provided additional APIs for messaging, data session control, charging, mobility management, querying device capabilities, and other features.

By 2000, BT and their associates decided to open the Parlay Group to participation from the industry, causing the group to grow significantly. Around the same time, the 3GPP had started specifying APIs to facilitate the concept of service roaming for 3G mobile networks. As the 3GPP APIs showed significant synergy with the Parlay APIs, the 3GPP and Parlay agreed to align their specifications in 2001. The 3GPP called the resulting API specifications open service access (OSA), and the 3GPP's regional standardization bodies elevated them to worldwide standards. In the remainder of this chapter, we'll refer to OSA as the aligned OSA and Parlay specifications.

5.3.2 Open Service Access

OSA not only defines network APIs, it also provides a framework for security and service management. The framework interfaces allow OSA to identify, authenticate, and authorize a third-party application before granting it access to the network APIs. An OSA service provider can set different privilege levels for different third-party applications, depending on their trust level. For example, the OSA framework can let some applications receive only call-related notifications while letting other applications also control calls through the API. OSA also supports nonrepudiation[13] by letting applications digitally sign an online agreement for the use of specific API features.

---

[12] Although the IT community uses lots of acronyms, Parlay is just a given name, not an acronym.
[13] Nonrepudiation refers to the mechanism to prevent a user from denying the actual use of a service (and refusing to pay).

OSA defines the framework and network APIs as separate interfaces and allows them to be provided by different enterprises. This allows a service provider to delegate the authentication, authorization, and nonrepudiation functions to an independent and specialized security firm or clearing house. Table 5.3 gives an overview of the network APIs defined in OSA. The numbers in the left-hand column refer to the 3GPP specification number and allow you to find the corresponding specification on the 3GPP website (3gpp.org). Table 5.3 omits auxiliary specifications such as the OSA overview (3GPP

Table 5.3
3GPP Open Service Access APIs

Specification	API	Allows a Third-Party Application To
29.198-03	Framework	Manage access to the OSA APIs. Authenticate mutually and digitally sign agreements (both the application and the platform) to use OSA features.
29.198-04-1	Call control common definitions	Set up, release, and manage calls, conferences, and multimedia connections. Receive notifications of call- and connection-related events.
29.198-04-2	Generic Call Control	
29.198-04-3	Multi Party Call Control	
29.198-04-4	Multi Media Call Control	
29.198-04-5	Conference Call Control	
29.198-05	User Interaction	Interact with a subscriber to play or display messages on his or her device and receive user input.
29.198-06	Mobility	Receive notifications of subscriber location and status.
29.198-07	Terminal Capabilities	Interrogate a terminal for its capabilities.
29.198-08	Data Session Control	Set up, release, and manage data sessions
29.198-10	Connectivity Manager	Negotiate and manage QoS and SLAs.
29.198-11	Account Management	Create, delete, and modify subscriber accounts.
29.198-12	Charging	Reserve and charge units of money, time, or data volume against a subscriber account (mainly for implementing prepaid communications services).
29.198-13	Policy Management	Create, update, delete, and retrieve policies, policy groups, and domains for policy-enabled services. Request the evaluation of a policy for a given service and subscriber.

Table 5.3
3GPP Open Service Access APIs *(Cont.)*

Specification	API	Allows a Third-Party Application To
29.198-14	Presence and Availability Management	Manage identities (subscribers) and agents (devices). Set presence and availability values for given identities and agents. Receive presence and availability events.
29.198-15	Multi Media Messaging	Send and receive messages short messages, multimedia messages, instant messages, and e-mails. Receive notifications about message delivery status.
29.198-16	Service Broker	Manage interaction between applications and services. Orchestrate the flow between the applications and services.

specification 29.198-01) and common data type definitions (3GPP specification 29.198-02).

The OSA specifications describe these APIs in OMG IDL (see Section 5.1.5) with the support of universal modelling language (UML) diagrams.

5.3.3 OSA Interface Structure

All OSA interfaces follow a more or less similar structure. For each interface, OSA defines two interfaces on the network side:

- `Ip<Interface>`[14] provides operations to control resources in the network. Here `<Interface>` is a placeholder for the name of the interface (e.g., `GenericMessaging` or `TerminalCapabilities`).
- `Ip<Interface>Manager` provides operations to create instances of classes that implement `Ip<Interface>` and request server-related event notifications such as overload conditions.

---

[14] All Parlay and OSA interfaces are prefixed with "Ip" (e.g., as in "IpConfCall"). This prefix is rather awkwardly chosen because it suggests some association with the Internet protocol (IP); however, in this context it stands for *interface Parlay*. Parlay and OSA use a notation scheme common in object-oriented programming: class names always begin with a capitalized string (e.g., IpCall) and operation names always begin with a lowercase string (e.g., routeReq). Names cannot contain spaces (e.g., "route Req" is not a valid operation name) and usually avoid the use of underscores (e.g., route_Req is a valid operation name but not recommended).

To allow the network to send notifications to the client application, the client application also has to provide two interfaces:

- `IpApp<Interface>` contains operations for communicating results and errors back to the client.
- `IpApp<Interface>Manager` contains operations to start and manage an instance of `IpApp<Interface>`.

In object-oriented terminology, these client-side interfaces are referred to as callback interfaces. Compared with the Observer pattern (Section 4.1.5), callback interfaces facilitate asynchronous communication between the client and the server in a more primitive way. They involve no subject and do not need a client to subscribe to notifications.

Figure 5.7 illustrates the general structure of the OSA interfaces. OSA doesn't follow this pattern rigidly for all interfaces, but most of the interfaces follow this structure.

Section 5.3.4 presents an example that describes the OSA call control API more in detail, so as to get an idea of the OSA API structure.

The complete cycle for using an OSA service consists of three phases:

- *Authentication:* before using OSA services, the application and the OSA framework authenticate each other. Authentication is part of the OSA framework interface and prevents unauthorized access.
- *Service selection:* once authenticated, the application selects the service interface to use (in this example, the call control interface). To ensure nonrepudiation, the OSA framework may request the signing of a service agreement before allowing the interface to be used.

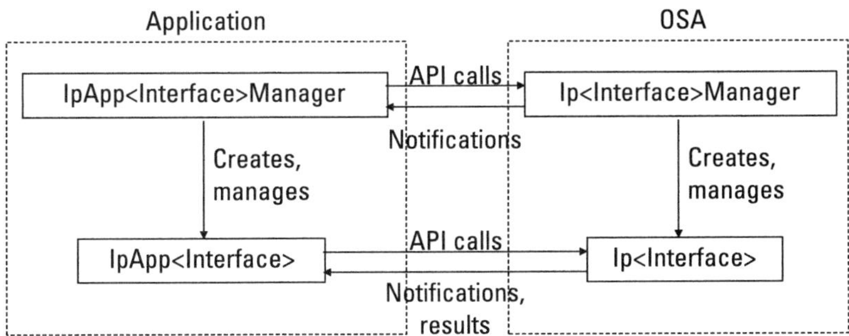

Figure 5.7   Common OSA interface structure.

- *Service use:* only after the authentication, service selection, and signing of the service agreement have been done, can the application start using the actual service.

Apart from providing security, the authentication and service selection process also allows OSA service providers to define permission profiles for different enterprise applications. The service provider can adjust the privileges it gives to each third-party application according to the level of trust it has.

For example, the OSA service provider can allow applications from a certain external entity to only receive notifications from the network (e.g., notifications of calls to a freephone number), while allowing more trusted entities to also establish calls.

5.3.4 Prepaid Calling: An Example

In this section, we'll see how a third-party application can manage a prepaid call using the OSA interfaces. Figure 5.8 shows the OSA interfaces involved and the flow of operations called on them.

Figure 5.8 describes the following steps:

1. The client application asks the service provider to notify it of a call attempt by a particular subscriber by calling the `enableCallNotification()` operation on the `IpCallControlManager` interface. This operation takes as a parameter a description of the event criteria to trigger the notification. In this case, the application will ask to be notified of a `P_EVENT_GCCS_ADDRESS_ANALYSED_EVENT` for the originating call from the subscriber. The application will also provide a (self)reference to the `IpAppCallControlManager` object that the server has to notify.

2. If at some point in time the subscriber picks up the phone and dials a number, this produces the `P_EVENT_GCCS_ADDRESS_ANAL-YSED_EVENT` the application subscribed to in step 1. The `IpCall-ControlManager` will first create a new `IpCall` object to represent this call. It then notifies the client application of the event by calling the `callEventNotify()` operation on the `IpAppCallCon-trolManager` interface. As a result, the call control manager on the application side creates an `IpAppCall` object to provide a notification interface for the new call.

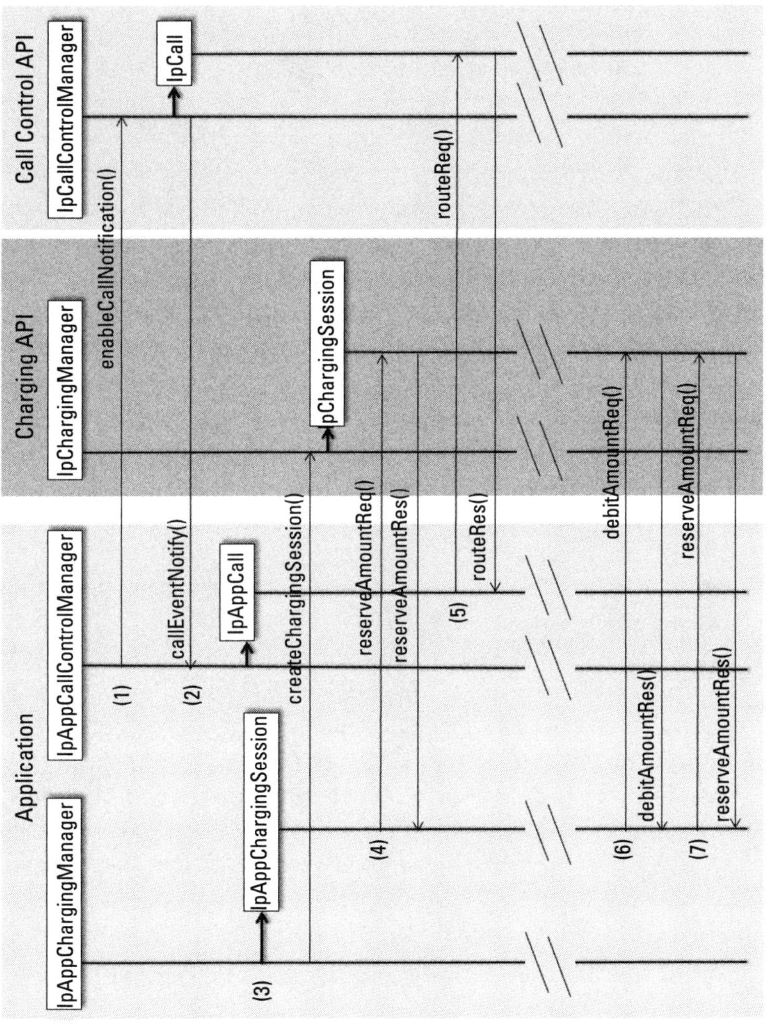

Figure 5.8 Prepaid call with the OSA API.

3. To access the subscriber's prepaid account, the client application must first start a charging session with the service provider. To do so, the application first creates a local `IpAppChargingSession` object and calls the `createChargingSession()` operation on the service provider's `IpChargingManager` interface. One of the parameters of this operation identifies the subscriber whose account to charge. This results in the creation of a new `IpChargingSession` object.

4. The application determines the cost of the call, and proceeds by reserving the amount for the first time unit of the call (e.g., one minute) by calling the `reserveAmountReq()` operation on the `IpChargingSession` object for this call created under point 3. The most important parameter of this operation is the amount to be reserved. If successful, the `IpChargingSession` confirms by calling the `reserveAmountRes()` operation on the application's `IpAppChargingSession` interface. This operation may carry a parameter specifying the maximum time during which the session will maintain the reservation.

5. After successful reservation of funds, the client application calls the `routeReq()` operation on the `IpCall` object to complete the call to its destination address. This operation carries the originating and terminating subscriber addresses as parameters. If the terminating party picks up, the `IpCall` confirms call completion by calling the `routeRes()` operation on the `IpAppCall` interface.

6. When the first unit of time expires, the client application will charge the reserved amount and then reserve a new amount for the next unit of time. It charges the reserved amount by calling the `debitAmountReq()` operation on the `IpChargingSession`, specifying the amount to be charged in the operation's parameters. The application will set the `closeReservation` parameter of this operation to `FALSE` to indicate that the reservation should continue in existence, even if its amount is 0 after the charge. The `IpChargingSession` will confirm a successful charge against the reservation by calling the `debitAmountRes()` operation on the application's `IpAppChargingSession` interface.

7. The application immediately makes a new reservation for the next time unit by repeating the operation calls described in step 4. If the reservation is successful, the application will let the call continue as is. Steps 6 and 7 will repeat for as long as the call continues, until either of the parties terminates the call. It is also possible that the subscriber's

account runs out of credit during the call. Figure 5.9 shows what happens in this case:

8. The client application makes the charge and tries to make a reservation for the next time unit, as described in step 7. If the subscriber has insufficient credit for the reservation, the `IpChargingSession` responds to the reservation request by calling the `reserveAmountErr()` operation on the application's `IpAppChargingSession` interface. This operation will carry the `P_CHS_ERR_NO_DEBIT` parameter telling the application that it cannot make any more charges against this account.

9. The application ends the call by calling the `release()` operation on the `IpCall` interface, which ceases to exist once the network has done the necessary signaling to release the call and its connections. The `ipCallControlManager` confirms the release of the call by calling `callEnded()` on the application's `IpAppCallManager` interface.

Of course, this example has a few simplifications. For example, one would expect the service provider to play a message to the subscriber, informing him or her of the lack of credit before releasing the call in step 9. Interacting with the call participants requires the additional OSA User Interaction API, omitted here to keep the example simple.

This examples demonstrates how the OSA interfaces follow a systematic object-oriented structure where each `Ip<Interface>` on the service provider side has a matching `IpApp<Interface>` on the application side and where all `req()` request operations have a matching `res()` or `err()` counterpart on the application side.

Almost immediately after their publication, the industry has criticized the 3GPP OSA and Parlay specifications for applying CORBA's synchronous communications model to an API that has a mostly asynchronous structure of communications. This manifests itself in the need for the application to implement the various `IpApp<Interface>` interfaces and their operations, so as to receive asynchronous responses from the service provider to their requests. Moreover, the 3GPP OSA uses strong typing, which produces well-defined data structures but also has a certain rigidity and makes communication between loosely coupled clients and servers more difficult. For this reason, the industry has been looking at defining the 3GPP OSA interfaces in a simpler, more asynchronous, and less rigid manner, using SOAP and REST.

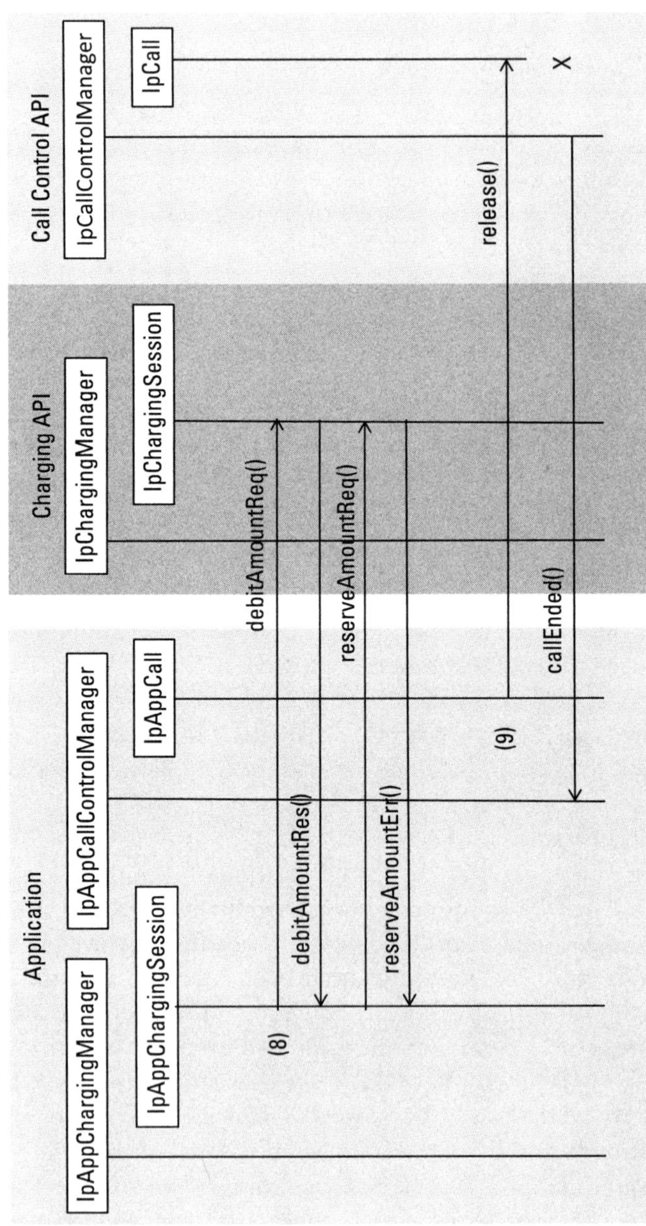

Figure 5.9   Subscriber out of credit.

5.4 Parlay X and Open Mobile Alliance

At the beginning of the millennium, WSs began gaining widespread adoption, causing the Parlay Group and the 3GPP to adapt the Parlay and OSA interfaces for use with SOAP and WSDL. This resulted in the specification of a new set of APIs called Parlay X.[15] Parlay X specifies network APIs as WSs, using WSDL as the specification language.

5.4.1 Parlay X

The Parlay Group and the 3GPP originally intended Parlay X to provide a WS API for the existing Parlay and OSA interfaces in Table 5.3 but eventually became a more independent API set. Some of the Parlay X APIs such as Short Messaging are simplified versions of OSA interfaces. But Parlay X also introduces new interfaces not supported by Parlay and OSA, such as interfaces for geocoding and policy management.

Table 5.4 gives an overview of the Parlay X specifications. Again this table omits the common data definitions in 3GPP specification 29.199-01.

The Parlay X APIs lack the framework interfaces specified in Parlay and rely on the WS framework for access and security management. Parlay X describes simpler and more up-to-date APIs as WSs, and has all but made obsolete the OSA specifications presented in Sections 5.3.2 and 5.3.3.

The Call Handling API allows an application to define simple call handling rules that resemble what can be specified with CPL (see Section 3.4.7). The Short Messaging API provides a stripped-down version of the Multimedia Messaging API, allowing for sending text messages only. The Parlay X Audio Call API corresponds to the OSA User Interaction interface of Table 5.3.

The Payment API corresponds with the OSA Charging interface, but the Parlay X Account Management API provides different functionality than the OSA interface with the same name.

The Parlay X Presence API is based on the XMPP and SIP SIMPLE presence protocols, and the IETF rich presence extensions to PIDF (see Section 3.6.1 and [9]). The data types for the Parlay X Presence API contain an extensive enumeration of activities, places, and moods.

The Parlay X Geocoding API provides additional information compared with the Terminal Location API. Whereas the latter provides the geographic coordinates of a device, the Geocoding API provides an address location (i.e.,

---

[15] Curiously, the 3GPP adopted the term OSA for the original Parlay specifications but kept the name Parlay X.

Table 5.4
Parlay X APIs

Specification	API	Allows a Third-Party Application To
29.199-02	Third Party Call	Set up or terminate a call. Add and delete call participants. Transfer a participant from one call to another.
29.199-03	Call Notification	Receive notification of call-related events (busy, no answer, answer), with the possibility to return information on how to proceed with the call. Receive input from the user as the result of a user interaction event in the call.
29.199-04	Short Messaging	Send and receive messages short messages (SMS). Receive notifications about message delivery status.
29.199-05	Multi Media Messaging	Send and receive messages short messages (SMS), multimedia messages (MMS), instant messages, e-mails, and other types of messages. Receive notifications about message delivery status.
29.199-06	Payment	Reserve and charge units of money, time, or data volume against a subscriber account (mainly for implementing prepaid communications services).
29.199-07	Account Management	Retrieve and recharge the balance of a prepaid account. Receive notifications about prepaid account-related events.
29.199-08	Terminal Status	Request the status (reachable, unreachable, or busy) of a device or a group of devices. Receive notifications about status changes for a device.
29.199-09	Terminal Location	Request one-time or periodic location information for a device or group of devices. Receive notifications about the change in location of a device or changes in the distance between two or more devices.
29.199-10	Call Handling	Set rules to handle incoming calls to a particular subscriber. The rules can specify white list or black list screening, conditional or unconditional call forwarding, or a combination of these.
29.199-11	Audio Call	Play text, audio, or video message to a subscriber and poll message status (playing, played). Get input (digits or audio) from a subscriber. Get and set media properties for the call and individual participants.
29.199-12	Multimedia Conference	Create and delete conference sessions. Invite and disconnect participants to and from a conference session. Get the properties of a conference session or the participants in the session.

Table 5.4
Parlay X APIs *(Cont.)*

Specification	API	Allows a Third-Party Application To
29.199-13	Address List Management	Create, delete, and query address lists. Set list ownership and permissions for list access. Add address to a list. Delete address from a list.
29.199-14	Presence	Get presence information for a subscriber. Receive presence updates. Publish presence information.
29.199-15	Message Broadcast	Send text messages to mobile and fixed devices in a given geographic area, specified as a circle or polygon. Receive notifications about broadcast delivery status.
29.199-16	Geocoding	Request the street address for a device or group of devices. Request the distance of a device from a given street address.
29.199-17	Appication Driven QoS	Set, modify, or remove a permanent or temporary QoS feature (e.g., uplink or downlink speed). Receive notifications about QoS-related events.
29.199-18	Device Capabilities and Configuration	Get the identifier (IMEI) of the device a subscriber is using. Receive notifications when a subscriber changes device. Get the capabilities of a device (returns an URL to a file that describes the capabilities). Get the list of configurations available for a given device. Select a configuration for a given device.
29.199-19	Multi Media Streaming Control	Start, pause, resume, transfer, and terminate media streams. Receive notifications about media stream status.
29.199-20	Multi Media Multicast Session Management	Create and delete multicast sessions. Request users to join or leave a multicast session. Get the list of multicast session participants. Receive notifications about users joining and leaving a multimedia session.
29.199-21	Content Management	Submit, modify, and delete content on a content server. Get metadata for content on the server, search content metadata for keywords. Receive notifications about the status of a content submission or modification (content providers). Receive notifications about new content submitted (subscribers).
29.199-22	Policy	Create, update, delete, and retrieve policies and domains for policy-enabled services. Request the evaluation of a policy for a given service and subscriber. Receive notification about changes to policies or policy domains.

a house number, street, city, region, or country, depending on the level of accuracy requested and available).

In 2009, the Parlay Group and the 3GPP agreed to transfer maintenance of OSA and Parlay X specifications to the Open Mobile Alliance (OMA). By that time, OMA had become a reference organization for specifying mobile application layer technologies, and many OMA members were also 3GPP and Parlay Group members. Since OMA has its own API specification program, it made sense to rationalize efforts and resources.

The 3GPP has stopped publishing updates of the OSA and Parlay X specifications from Release 10 onward. OMA has since then specified REST APIs based on the Parlay X specifications. The resulting set of APIs is called the OMA RESTful network APIs, which we'll discuss in the next section.

5.4.2 OMA RESTful Network APIs

OMA provides REST bindings for most, but not all, of the Parlay X APIs listed in Section 5.4.1. OMA provides RESTful APIs for the Parlay X interfaces for Address List Management. Audio Call, Call Notification, Device Capabilities, Multimedia Messaging, Payment, Presence, Third Party Call, Terminal Location, and Terminal Status. OMA provides additional API specifications as listed in Table 5.5.

Table 5.5
OMA RESTful Network APIs in Addition to Parlay X

API	Allows a Third-Party Application To
Anonymous Customer Reference Management	Create and delete an anonymous customer reference (ACR) for a particular subscriber and application. Retrieve the ACR for a particular subscriber.
Authorization Framework for Network APIs	Request an access token for a protected resource (e.g., one of the other network APIs). Access the protected resource with the token.
Capability Discovery	Retrieve the own service capabilities, or the service capabilities of another subscriber. Create or remove sets of service capabilities for a given device, called service capability sources. Create, update, or remove service capabilities.
Chat	Create, terminate and accept chat sessions. Send a chat message. Set and get the status of a chat message (sent, delivered, displayed, failed). Add or remove participants to or from a group chat session. Receive notifications about chat-related events (invitations to chat sessions, incoming chat messages, status of sent messages).

Table 5.5
OMA RESTful Network APIs in Addition to Parlay X *(Cont.)*

API	Allows a Third-Party Application To
Converged Address Book	Create, delete, and retrieve contacts and contact lists for a subscriber. Create, delete, and retrieve contacts in a contact list. Create, delete, and retrieve attributes for contacts and contact lists. Receive notifications about address book changes for a given subscriber.
Customer Profile	Retrieve the customer profile attributes for a given customer.
File Transfer Image Share Video Share	Create and delete file transfer, image sharing, or video sharing sessions. Retrieve session information. Receive notifications about session-related events (invitation to session, acceptance or decline of session, file transfer or image/video sharing successful or failed, session end).
Notification Channel	Create and delete a notification channel for events. Retrieve notification channel information. Retrieve events from the channel through long polling.[1]
Push	Create, update, and cancel a push message. Query the status of a push message. Receive notifications about push message delivery status.
Roaming Provisioning	Create and retrieve a roaming subscription for a given subscriber, and update its status (pending, active, deactivated, canceled, and so on). Receive notifications about changes to roaming subscriptions.
Service User Profile Management	Create, delete, and retrieve service user profiles. Create, delete, and retrieve attributes (name-value pairs) in a service user profile.
WebRTC Signaling	Create, terminate, accept, reject, and cancel WebRTC audio or video sessions. Send, answer, and update an SDP offer for WebRTC session parameters. Receive notifications about events related to WebRTC sessions.

[1] In a normal polling procedure, the client queries the server periodically and the server responds immediately regardless of whether it has new information. Long polling is a variation of polling whereby the server does not respond to the client until it has some result to return. This has several advantages: it reduces traffic and does not require the client to send periodic requests to the server.

The anonymous customer reference (ACR) management API lets an application create an anonymous reference for a particular subscriber. This allows the application to make anonymous a subscriber identity so it can make use of network services or APIs without having to share the subscriber's real network identity (MS-ISDN). Regulation sometimes requires such protection of the subscriber's privacy.

ACRs have a limited lifetime and expire after a set time. The ACR management API lets the application query the status of an ACR and refresh an ACR that is about to expire.

The Authorization Framework for Network APIs are based on IETF OAuth 2.0 specifications in RFC6749 [10], RFC6750 [11], and RFC7009 [12].

The purpose of the Capability Discovery API resembles that of the Device Capabilities API of Parlay X (see Section 5.4.1), but it has a different structure. The Device Capabilities API provides operations to configure the device, while the Capability Discovery API only allows an application to query the capabilities of the devices used by a subscriber.

The Chat API allows an application to manage one-to-one chat and group chat, and expand a one-to-one chat session into a group session. Applications can also use the Chat API to send messages to each other directly in sessionless one-to-one chat.

Many operators now build their communication services around a contact list or address book concept. Storing a subscriber's contacts in the network rather than in the device gives a service provider more control over a subscriber's communications and makes it less likely that the subscriber will use OTT applications (see Section 3.4). For this reason, many consider the Converged Address Book one of the OMA RESTful network APIs of the highest strategic value.

The File Transfer, Video Share, and Image Share APIs have a very similar structure, both in terms of resources and HTTP operations enabled for these resources. These APIs let a client application create a session for file transfer, video sharing, or image sharing; retrieve session information; and subscribe to session-related events. The session itself contains the necessary information about the source and target of the transfer or share and the file(s) to be transferred or shared. The client application can receive session events (invitation, acceptance, end) and notifications of the success or failure of the file transfer or share.

The Notification Channel API allows a client application to create notification channels for events from other OMA enablers. An application can use the Notification Channel API in push or pull mode. When used in pull mode, the application retrieves events from the channel by long polling, which uses the HTTP POST rather than the HTTP GET method. When used in push mode, the application will receive the events through the OMA Push enabler [13, 14], an evolution of the WAP Push protocol that does not form part of this API.

OMA specified the Roaming Provisioning API to help service providers comply with European legislation on roaming EU531_2012 [15]. This regulation requires service providers to decouple roaming as a service from domestic communications subscriptions and entitles subscribers to contract roaming services from Alternative Roaming Providers (ARP) who are different from their Domestic Service Provider (DSP).

The Roaming Provisioning API lets an ARP create a roaming subscription with the DSP for a given subscriber, as identified by the IMSI. It also allows the DSP to notify the ARP of changes to the subscription. Once created, the ARP and the DSP can activate, deactivate, suspend, or cancel the roaming subscription.

The OMA Push REST API provides a RESTful interface for the OMA push access protocol (PAP) [13]. It makes extensive reuse of the PAP data structures but replaces the PAP messages with RESTful HTTP messages.

The WebRTC Signaling API allows an application to manage WebRTC sessions. WebRTC is a joint IETF and W3C standard that allows web browsers to act as end points for audio or video sessions. Like SIP, WebRTC uses SDP to negotiate session parameters between the end points, using an offer-answer-update dialog.

OMA has also published guidelines that help authors define new RESTful network API specifications in a consistent style [16]. Although these guidelines are meant for OMA members that specify APIs, they also give developers more insight into how OMA structures its RESTful API specifications.

The OMA specifications provide machine-readable supplementary files that contain XML schema definitions of the data types used in the API. OMA has machine validated these XML schemas prior to publication.

5.4.3 Prepaid Calling: An Example

In this section we'll revisit the prepaid calling example of Section 5.3.4, but this time implemented with Parlay X APIs. Unlike OSA, the simpler Parlay X call control API does not have any interface to notify an application of an outgoing call attempt by a subscriber and subsequent control of that call. Parlay X only lets a client application control calls it has created itself. We must therefore assume that the network somehow informs the application that the subscriber wants to make a prepaid call. There are several ways to achieve this, which we'll not discuss in detail here, such as having the subscriber's device send an unstructured supplementary service data (USSD)[16] message, clicking a link on a website, or calling a special number.

---

[16] USSD provides a real-time low-bandwidth data channel between a mobile device and a server that the operator can use to implement mobile applications and services. The USSD format resembles that of SMS, but USSD uses a data connection rather than a store-forward mechanism, making it more suitable for real-time interactions with the user.

Figure 5.10 shows the flow from the moment that the network has made the application aware of the subscriber's intention to make a prepaid call to a given number. All interactions in Figure 5.10 are WS calls where HTTP POST messages carry requests and HTTP response carry results.

Figure 5.10 presents the following steps:

1. Once the client application knows that the subscriber wants to make a prepaid call to a given number, it determines what the cost per time unit to the destination number is and sends a HTTP POST message to the Payment API with a SOAP request to reserve the corresponding amount. The SOAP request has the following body (name spaces omitted for simplicity):

```
<Envelope>
  <Body>
    <ReserveAmountRequest>
      <EndUserIdentifier>
          tel:+34696858585
      </EndUserIdentifier>
      <Charge>
          <ChargingInformation>
              <Description>
```

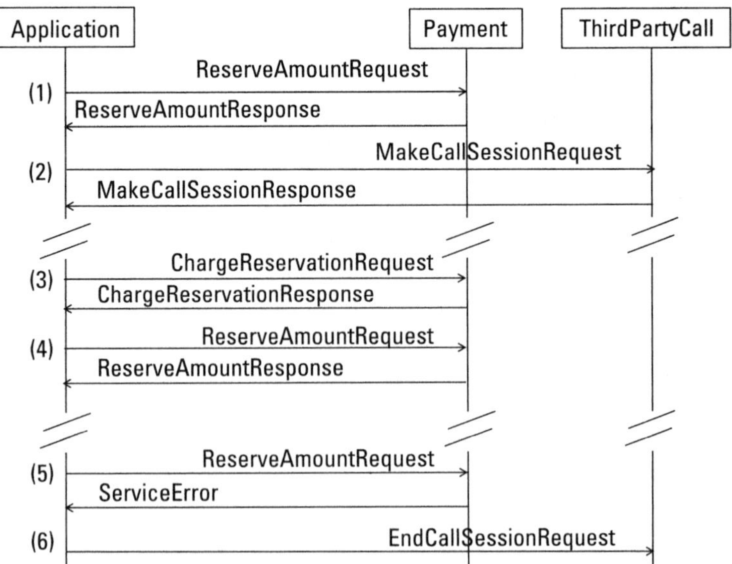

Figure 5.10 Prepaid call with the Parlay X (SOAP) API.

```
             Prepaid call to number +34696904123
         </Description>
         <Amount>
            0.25
         </Amount>
       </ChargingInformation>
     </Charge>
   </ReserveAmountRequest>
 </Body>
</Envelope>
```

If the subscriber has sufficient credit, the server reserves the amount on the subscriber's account so that no other application can charge it. It responds to the client application with a HTTP 200 OK response, confirming the reservation in its SOAP payload (the result string is some unique reference for the reservation):

```
<Envelope>
  <Body>
    <ReserveAmountResponse>
      <Result>
         876498Hhif98y578
      </Result>
    </ReserveAmountResponse>
  </Body>
</Envelope>
```

2. The application instructs the ThirdPartyCall API to set up the call from the subscriber to the destination number by sending a HTTP POST message with the following SOAP body:

```
<Envelope>
  <Body>
    <MakeCallSessionRequest>
      <CallParticipants>
        tel:+34696858585, tel:+34696904123
      </CallParticipants>
    </MakeCallSessionRequest>
  </Body>
</Envelope>
```

If the terminating party picks up the phone, the server responds with a `HTTP 200 OK` response containing the following confirmation, where `Hu957i3cuyf48e4w` is a unique identifier for the call session:

```
<Envelope>
  <Body>
    <MakeCallSessionResponse>
      <CallSessionIdentifier>
        Hu957i3cuyf48e4w
      </CallSessionIdentifier>
    </MakeCallSessionResponse>
  </Body>
</Envelope>
```

3. After a unit of time (e.g., a minute) has expired, the application instructs the Payment API to charge the reserved amount by sending a `HTTP POST` message with the following SOAP body (note that `876498Hhif98y578` is the reference of the reservation returned by the server under step 1):

```
<Envelope>
  <Body>
    <ChargeReservationRequest>
      <ReservationIdentifier>
        876498Hhif98y578
      </ReservationIdentifier>
      <Charge>
        <ChargingInformation>
          <Description>
            Prepaid call to number +34696904123
          </Description>
          <Amount>
            0.25
          </Amount>
        </ChargingInformation>
      </Charge>
      <ReferenceCode>
        fh74hf30Br5C0rC4r7v8
      </ReferenceCode>
    </ChargeReservationRequest>
  </Body>
</Envelope>
```

4. The application immediately reserves a new amount to pay for the next unit of time, using the same SOAP requests and responses as under step 1. If the Payment API confirms the reservation, the application lets the call continue as is.

5. Steps 3 and 4 repeat throughout the duration of the call, until one of the parties terminates the call. It is also possible that the subscriber runs out of credit during the call. In this case the Payment API will respond with `HTTP 400 Bad Request` error, in which it provides a service error code specified by Parlay X.

6. The application then instructs the Third Party Call API to end the call session by sending a `HTTP POST` message with the following SOAP body (note that `Hu957i3cuyf48e4w` is the call session identifier returned by the server under step 2):

```
<Envelope>
  <Body>
    <EndCallSessionRequest>
      <CallSessionIdentifier>
        Hu957i3cuyf48e4w
      </CallSessionIdentifier>
    </EndCallSessionRequest>
  </Body>
</Envelope>
```

Of course, this flow has a somewhat simplified structure. Before executing step 6, the application would typically play an announcement to the subscriber saying that his or her credit has run out. This involves an additional Parlay X component, the Audio Call API. We've omitted this for simplicity.

5.5 GSMA OneAPI

OMA published specific RESTful APIs for RCS. These are not new APIs but profiles[17] of the Notification Channel, Address Book, Presence, Messaging, Chat, File Transfer, Third Party Call, Call Notification, Video Share, and Image Share APIs described in Section 5.4.2. For its part, GSMA has launched the OneAPI initiative, which publishes a number of RESTful APIs derived from the OMA RESTful network specifications of the Parlay X based network APIs.

---

[17] An API profile specifies a subset of the operations provided by the original API. A profile can make mandatory operations optional and vice versa, and may have conditional definitions such as "if operations X and Y are supported, then operation Z must also be supported."

5.5.1 OneAPI Interfaces

Although mostly based on the RESTful network APIs specified by OMA, GSMA has published OneAPI on their web site as specifications in their own right. In most cases, the OneAPI specifications simplify the OMA specifications. Table 5.6 shows the OneAPI interfaces and their relationship with OMA RESTful network APIs.

Table 5.6
GSMA OneAPI

OneAPI	Based On	Allows an Application To
Voice Call Control	Third Party Call and Audio Call—OMA RESTful network API	Set up or terminate a voice call. Add and delete call participants. Play an audio message to call participants. Get keypad input from call participants. Receive notifications of call-related events and keypad input events.
Device Capability	Device Capabilities and Configuration—OMA RESTful network API	Get the identifier (IMEI) of the device a subscriber is using. Receive notifications when a subscriber changes device. Get the capabilities of a device (returns an URL to a file that describes the capabilities).
Data Connection Profile	N/A	Get the connection and roaming status of subscriber's device.
SMS	Short Messaging—OMA RESTful network API	Send and receive messages short messages (SMS). Receive notifications about message delivery status.
Payment	Payment—OMA RESTful network API	Charge a subscriber, refund a subscriber, or reserve an amount against a subscriber account. Query the status of a payment transaction.
MMS	Multimedia Messaging—OMA RESTful network API	Send and receive messages multimedia messages (MMS). Receive notifications about message delivery status.
Location	Terminal Location—OMA RESTful network API	Retrieve the location of one or more devices with a given accuracy
Anonymous Customer Reference	Anonymous Customer Reference—OMA RESTful network API	Create an anonymous customer reference (ACR) for a particular subscriber. Query the status of the ACR for a particular subscriber.

The OneAPI Voice Call Control API simplifies and combines the OMA Third Party Call and Audio Call RESTful network APIs. Only the Data Connection Profile API does not really have its counterpart in the OMA RESTful network specifications. Whereas the OMA APIs use XML schema to define the data structures, OneAPI allows the use of either XML or JSON.[10]

By now you probably start losing oversight of the different specifications and their relationships. Figure 5.11 shows how Parlay, Parlay X, OMA RESTful network APIs, and GSMA OneAPI relate to each other.

The APIs published by both OMA and GSMA draw upon Parlay X but use REST rather than SOAP for object access. At the surface, Parlay X and OneAPI do not exhibit great differences, as both use HTTP messages. But as we've seen in Section 5.2.5, the way in which they use HTTP differs. To appreciate this difference, let's return to prepaid calling example of Section 5.4.3 and see how OneAPI interfaces enable this application.

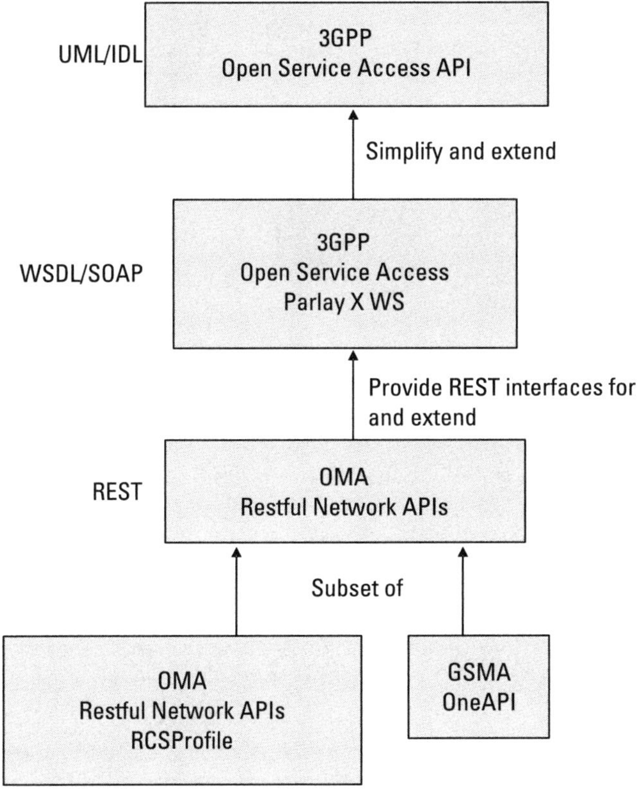

Figure 5.11　Relationship between Parlay, Parlay X, OMA APIs, and OneAPI.

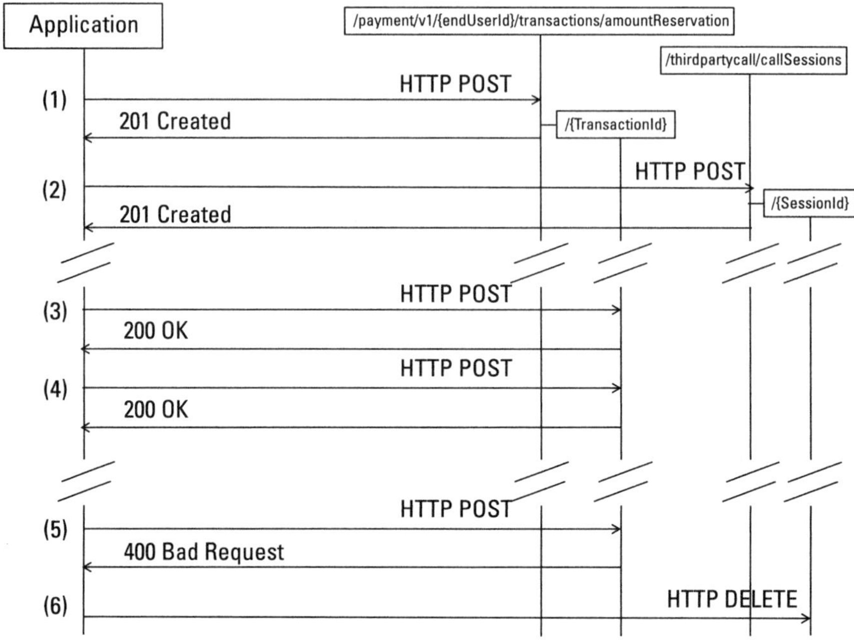

Figure 5.12   Prepaid call with OneAPI (REST).

5.5.2   Prepaid Calling: An Example

Figure 5.12 shows the flow of HTTP messages for a third-party controlled prepaid call application, using OneAPI interfaces. Unlike Figure 5.10, Figure 5.12 explicitly shows the HTTP messages exchanged. This is because with REST, the HTTP request type defines the semantics of the message, unlike SOAP where the message payload carries all information.

The initial conditions are the same as for Parlay X. OneAPI does not allow an application to be notified of an outgoing call attempt, so we assume that the application has a way of knowing that a subscriber tries to make a call to a given number.

The HTTP message flow is the following:

1. Once the client application knows that the subscriber wants to make a prepaid call to a given number, it determines what the cost per time unit to the destination number is and sends a `HTTP POST` request to the URL `http://serviceproviderXYZ.com/payment/v1/tel%3A%2B34696858585/transactions/amountReserva-tion`, where `serviceproviderXYZ.com` is the domain of the

service provider providing the Payment API and +34696858585 is the identity of the subscriber trying to make the prepaid call. The JSON payload of this HTTP message is

```
{"amountReservationTransaction": {
    "endUserId": "tel:+34696858585",
    "paymentAmount": {"chargingInformation": {
        "amount": "0.25",
        "description": "Prepaid call to number
                        +34696904123"
    }},
    "referenceSequence": "1",
    "transactionOperationStatus": "Reserved"
}}
```

If successful, the server will provide a HTTP 201 Created response, returning the URL of the reservation in the Location header line:

```
Location: http://serviceproviderXYZ.com/
payment/v1/tel%3A%2B34696858585/transactions/
amountReservation/r0001
```

The reply will also echo the JSON payload of the request to confirm the amount reserved.

2. The application proceeds to set up the call by sending HTTP POST request to the URL http://serviceproviderXYZ.com/third-partycall/callSessions. This HTTP request has the following JSON payload to specify the call participants:

```
{"callSessionInformation": {
    "participant": [
    {"participantAddress": "tel:+34696858585"},
    {"participantAddress": "tel:+34696904123"}
    ]
}}
```

If successful, the server will provide a HTTP 201 Created response, returning the URL of the reservation in the Location header line:

```
Location: http://serviceproviderXYZ.com/
thirdpartycall/callSessions/c0001
```

The reply also contains a JSON payload to confirm the status of the call and its connections, and to provide the URLs that allow the application to manage the participants in the call:

```
{"callSessionInformation": {
    "participant": [
     {"participantAddress": "tel:+34696858585",
      "participantStatus":
"CallParticipantConnected",
      "resourceURL": "http://serviceproviderXYZ.
com/thirdpartycall/callSessions/c0001/
participants/p001"
     },
     {"participantAddress": "tel:+34696904123",
      "participantStatus":
"CallParticipantConnected",
      "resourceURL": "http://serviceproviderXYZ.
com/thirdpartycall/callSessions/c0001/
participants/p002"
     }
    ],
    "resourceURL": "http://serviceproviderXYZ.com/
thirdpartycall/callSessions /c0001",
    "terminated": "false"
}}
```

3. After a unit of time (e.g., a minute) has expired, the application instructs the Payment API to charge the reserved amount by sending a HTTP POST message to the URL `http://serviceproviderXYZ. com/payment/v1/ tel%3A%2B34696858585/transactions/ amountReservation/r0001` with the same JSON payload as under step 1 except with `referenceSequence` set to 2, as this is a new transaction, and `transactionOperationStatus` set to `Charged` to indicate the desired status. If the charge is successful, the server replies with a HTTP `200 OK` response message containing the following JSON payload:

```
{"amountReservationTransaction": {
    "endUserId": "tel:+34696858585",
    "paymentAmount": {
```

```
        "totalAmountCharged": "0.25"
        "amountReserved": "0",
        "chargingInformation": {
            "amount": "0.25",
            "description": "Prepaid call to number
                        +34696904123"
        },
    },
    "referenceSequence": "2",
    "resourceURL": "http://serviceproviderXYZ.com/
payment/v1/ tel%3A%2B34696858585/transactions/
amountReservation/r0001",
    "transactionOperationStatus": "Charged"
}}
```

This JSON description is similar to the description of the reservation transaction under step 1, but it has a few additional elements: `transactionOperationStatus` confirms that the server has made a successful charge, `totalAmountCharged` states the amount charged against the reservation, and `amountReserved` shows the amount remaining in the reservation after the charge. In this case `amountReserved` is 0 because the entire reservation has been charged.

4. The application reserves a new amount for an additional unit of call time. It does so by making a `HTTP POST` request to the URL of the reservation `http://serviceproviderXYZ.com/payment/v1/ tel%3A%2B34696858585/transactions/amountReservation/r0001`. This request has a JSON payload similar to that of the initial reservation request under step 1, with `referenceSequence` updated to 3. It may omit `description` and `endUserId` since the server already knows their values. If successful, the server replies with a `HTTP 200 OK` reply, which carries the following JSON payload:

```
{ "amountReservationTransaction": {
    "endUserId": "tel:+34696858585",
    "paymentAmount": {
        "totalAmountCharged": "0"
        "amountReserved": "0.25",
        "chargingInformation": {
            "amount": "0.25",
            "description": "Prepaid call to number
```

```
                                    +34696904123"
            },
        },
        "referenceSequence": "3",
        "resourceURL": "http://serviceproviderXYZ.com/
    payment/v1/ tel%3A%2B34696858585/transactions/
    amountReservation/r0001",
        "transactionOperationStatus": "Reserved"
}}
```

This JSON response bears strong similarity to the response sent back for a charging request under step 3, except that `transactionOperationStatus` has value `Reserved` to confirm the reservation, `amountReserved` shows the total amount reserved, and `totalAmountCharged` has value 0 to show that nothing has been charged.

5. Steps 3 and 4 repeat for the duration of the call, until one of the parties terminates the call. It is also possible that the subscriber runs out of credit during the call. In this case, the server can't make any more reservations and will return a `HTTP 400 Bad Request` reply.

6. If the application is not able to make any more reservations, it will proceed to end the call by sending a `HTTP DELETE` request to the URL of the call: `http://serviceproviderXYZ.com/third-partycall/callSessions/c0001`. This request does not need a JSON payload. The server will reply with `HTTP 200 OK` providing a summary of the aborted call in its JSON payload.

Although this flow shows clear resemblance with the SOAP based Parlay X exchange in Ssection 5.4.3, it also demonstrates how REST uses HTTP requests and URLs in a more explicit way to request specific actions on web resources. These examples also show the use of SOAP (XML) and JSON to describe requests and objects. In practice, both are used in web services today, with JSON gaining popularity in simpler applications because of its lean syntax.

5.6 Device APIs

In the previous sections, we've only considered network-side APIs (i.e., interfaces to network functions exposed to third-party applications by service providers). OMA has published APIs that give applications running in a device access to device managed connectivity, identity, and security functions.

Device APIs introduce new challenges because they depend on the device's operating system, and operating system vendors are notoriously reluctant to let third parties control and manage the device. They have several reasons for this, such as maintaining security and device integrity, but also the more strategic aim to maintain exclusive control over the user experience. The fact that neither Apple nor Google are members of OMA probably speaks for itself.

Device APIs therefore have more questionable viability and OMA, the 3GPP, and GSMA have specified only a handful of device APIs, far less than network APIs.

Since device APIs are device-internal, specifying them with SOAP or REST brings unnecessary complexity and overhead. OMA specifies some of its device APIs with WebIDL,[18] a W3C adaptation of IDL (Section 5.1.5). We'll look at four specific device APIs in more detail in the following sections.

5.6.1 SIM Application Toolkit

Originating in the mid-1990s, the SIM application toolkit (SAT)[19] represents the most veteran of device-based APIs. SAT differs somewhat from the other APIs we've seen in this chapter not only because it's a device API, but also because its main purpose is to support applications installed on the SIM by the operator, rather than third-party applications.

The 3GPP's SAT specifications [17, 18] emerge from the device-SIM partition discussed in Section 2.7.3. Whereas a mobile device has the functions to connect to mobile networks and provides the user interface (microphone, speaker, keyboard, screen), the SIM holds the essential subscription data and performs the required security protocols for network access.

Being a smart card with a processor, a SIM can run (small) applications. The operator has control over what it installs and runs on its SIMs but not

---

[18] The World Wide Web Consortium (W3C) defined WebIDL as an adaptation of the Object Management Group (OMG) IDL (Section 5.1.5), intended for use in web-based APIs. WebIDL is very similar to OMG IDL but has some features that improve compatibility with ECMAScript, the ISO language standard that forms the basis for Javascript. For example, WebIDL redefines some of the IDL data types so they better match the ECMAScript data types.

[19] The USIM application for 3G and 4G networks has its own USIM application toolkit (USAT), which is largely compatible with SAT. For simplicity, we'll use only the terms SIM and SAT in this section to refer also to USIM and USAT, respectively.

necessarily on what goes on in the rest of the device, which is at the discretion of the device manufacturer and the device owner. The SIM provides operators with a controlled environment to install and run their applications and services, which cannot easily be tampered with.

But a SIM-based application does expect a standard interface to the mobile device so it can access the device's functions and peripherals such as the screen, keyboard, data connections, GPS receiver, and so on. The SAT provides this standard interface between the SIM and the device, as shown in Figure 5.13. Some of the commands a SIM-based application can make to a mobile device that supports SAT are open (data) channel, send data, set up call, send short message, send USSD, open browser, play tone, and create menu.

SAT also lets SIM-based applications process outgoing calls and short messages before the device sends them to the network. This allows a SIM application to change the destination address of the message or call and even to turn a short message into an USSD or a call.

One typical SAT application lets a prepaid subscriber top up his or her prepaid account through a menu on the device's screen, and sends the requested amount and payment details to the operator's prepaid account server by USSD or data connection.

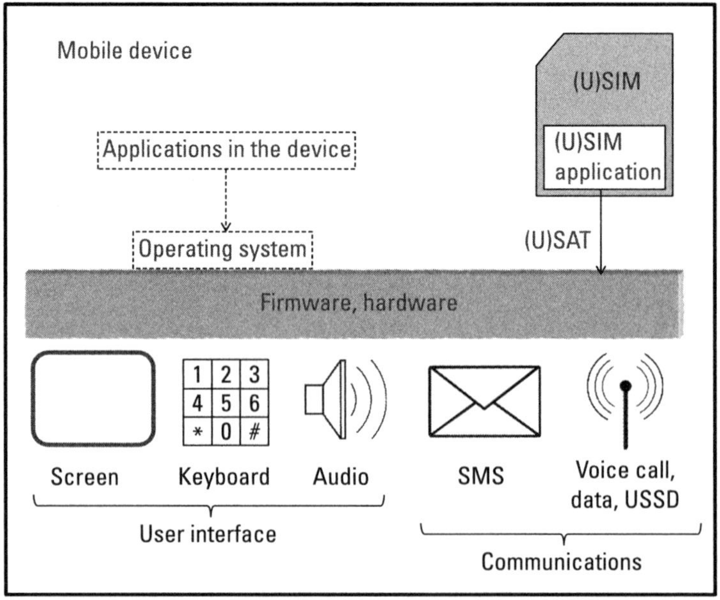

Figure 5.13   SIM application toolkit.

The SAT interface between SIM and device has a low-level binary message format without any binding to a particular programming language. However, one can still consider SAT an API because it lets SIM application programmers use well-defined interfaces to the device.

5.6.2   Mobile Execution Environment

The mobile execution environment (MExE), first specified by the 3GPP in 2000 and maintained to date [19], defines a standard set of run time resources in a mobile device. With MExE, operators aimed for better interoperability of applications and a more consistent user experience on different devices.

At the time the 3GPP specified the first MExE standard in the late 1990s, mobile data bandwidth was too narrow to use HTTP, and mobile devices did not have sufficient capacity to display standard HTML pages. Instead they used the wireless application protocol (WAP) to give a WWW-like experience in mobile devices. WAP is a standard from the WAP Forum (merged into the Open Mobile Alliance in 2002) consisting of

- *Wireless session protocol* (WSP), *wireless transaction protocol* (WTP), and *wireless datagram protocol* (WDP): a suite of lightweight binary protocols for sending low-volume text and graphics to mobile terminals. Note that these protocols bear no relationship with TCP/IP and do not provide Internet connections in a mobile device. They have the sole purpose of delivering WAP content to mobile devices over low bandwidth raw data links.
- *Wireless markup language* (WML): a simple markup language for use in resource-constrained devices. Unlike HTML, which describes pages, WML uses a card deck paradigm to structure concise hypertext content with low-resolution graphics.
- *WMLscript:* a scripting language that provides a stripped down version of Javascript for use with WML cards.

Also at the end of the 1990s, manufacturers began building mobile devices with some processing capacity, most of them supporting Java as programming language. WAP and Java formed an odd couple in mobile devices, because WAP can download Java applets but does not provide an execution environment for them, and WMLscript and Java can end up competing for resources in the device.

It proved difficult for operators to give a consistent look and feel to their applications because different devices have distinct capabilities such as screen size, colors and resolution, text input, audio, and processing power. They conceived MExE to provide a consistent set of run time resources for mobile devices that allows applications to have more consistent and predictable behavior on different devices.

MExE draws on existing standards and specifies how they should be used and combined. Doing so, MExE defines four compatibility categories for devices, called classmarks:

- *Classmark 1* devices provide WAP functionality (i.e., WML, WMLScript, and WSP/WTP/WDP).
- *Classmark 2* devices support the Personal Java environment extended with the telephony-specific JavaPhone API. Personal Java and JavaPhone are predecessors of JME (see Section 4.2.3) from the late 1990s. JavaPhone allows an application on the terminal to send and receive short messages (SMS), set up and answer calls, and monitor and control the power consumption of the device.
- *Classmark 3* devices support JME, in particular CLDC and MIDP (see Section 4.2.3).
- *Classmark 4* devices support the common language infrastructure (CLI) with the compact profile (CP). One can regard CLI as Microsoft's counterpart of the Java run time environment and CP as the equivalent of CMIP on CLI. CLI is now an ISO standard [20] and provides byte code compilers for several languages including Visual Basic and C#.

Note that classmarks define feature sets that have no upward compatibility (e.g., a classmark 4 terminal does not necessarily support classmark 3, 2, or 1 features). A device must support at least one classmark in order to have MExE compatibility and may support any combination of classmarks.

Apart from these classmarks, MExE defines a collection of common functions, including

- *Capacity negotiation:* devices that support MExE use the W3C composite capability/preferences profiles (CC/PP) format [21] with the attributes specified in the OMA UAProf standard [12, 22] to negotiate their capabilities with servers in the network. This allows an application server in the network to adapt its content to the classmark and capabilities (screen size, screen resolution, and so on) of a particular device.
- *Security:* MExE specifies procedures for permission and certificate management, defining what parties (operator, device manufacturer,

subscriber, third party) have access to which resources on the device and how to download, store, and validate certificates and encryption keys.

- *Quality of service API:* this API allows applications in the device to manage QoS at a high level without having any knowledge of the underlying connection.

5.6.3 Device Management Client API

The device management (DM) specifications are among the most important publications of OMA. DM allows a service provider to diagnose and configure a subscriber's device remotely. Service providers around the world use DM on a broad scale.

DM follows a specific model whereby managed objects (MOs) represent the configuration of a device. DM allows a service provider to manage these MOs. DM requires a server in the network and a client application on the device.

The DM client API allows applications on the device to interact with the DM client. More specifically, the DM client API lets an application create and retrieve MOs and set their attributes. The DM API uses the Observer pattern described in Section 4.1.5. An application can register as a listener (observer)

Table 5.7
Main Functions of the DM Client API

Object	Function	Description
DMClient	startDMSession()	Starts a DM session between client and server
	createMO()	Creates a new MO
	getMOByURI()	Retrieves the MO at the specified URI
MO	getNodeValue()	Gets the value of a node (attribute) of an MO
	setNodeValue()	Sets the value of a node for an MO
DM Subject	addDMUpdateListener()	Registers an application as a listener for DM events
	removeDMUpdateListener()	Deregisters an application as listener for DM events
Application	handleDMUpdate()	Reports an event to a listener

for an MO and receive updates when the state of the MO changes. Table 5.7 gives an informal list of the most important functions of the API.

5.6.4 Open Connection Manager API

The OMA open connection manager API (OpenCMAPI) gives applications access to a broad range of device features, including the data connectivity the device supports. Despite its name, the OpenCMAPI has a broader scope than just the management of connections in a device. It also provides functions to manage interactions with the SIM[20] (such as changing the PIN and PUK), with the positioning system (GPS), and with communications services such as short messaging and push.

The OpenCMAPI has a less explicitly object-oriented structure than the DM client API described in the previous section. It simply provides an extensive list of functions an application can call on the API. The OpenCMAPI specification structures the functions in several groups:

- *Discovery:* detect and manage communications ports supported by the device, such as USB, infrared, Bluetooth, and so on;
- *Device services:* get information about the hardware, firmware, and identifiers (IMEI, IMSI, MS-ISDN) and manage the PIN on the SIM;
- *Network:* Get information about the home and serving networks of a subscriber;
- *Network connection services:* manage cellular network connectivity: detect available networks, select a preferred network, and connect to network;
- *WLAN:* manage WLAN connectivity—scan networks, select a preferred network, and connect to a network;
- *GNSS:* manage the global navigation satellite system (GNSS)[21] receiver and request device location;
- *SMS:* send and receive short messages.

---

[20] Whereas the term SIM used to be employed universally (and often still is), standards organizations 3GPP and ETSI now distinguish between the smart card and the application running on it. They refer to the card (hardware) as the universal integrated circuit card (UICC), running the SIM (for GMS) or USIM (for UMTS and LTE) application.

[21] GNSS is the general name for satellite-based navigation systems. Most mobile devices currently use the global positioning system (GPS) provided by the U.S. Navstar system, but Russia (GLONASS), China (COMPASS), and Europe (Galileo) have developed their own systems, which are at different stages of deployment and will be supported by mobile devices in the near future.

Table 5.8 lists some of the key functions of the OpenCM API. The table does not provide an exhaustive listing, but it gives an idea of the scope of this API.

Table 5.8 shows that the OpenCMAPI is a rich API with a broad range of functions. Apart from the functions listed in this table, the Open CMAPI also allows an application to start a DM session and set certain DM parameters such as the firmware update feature. In this respect, the OpenCMAPI partly overlaps with the DM client API we've seen in the previous section, although it doesn't support all DM features.

Table 5.8
A Selection of OpenCM API Functions

Function	Description
CMAPI_API_Open()	Start the open CM API
CMAPI_Discovery_DetectDevices()	Detect communications ports (USB, infrared, Bluetooth, and so on)
CMAPI_Discovery_OpenDevice()	Open a communications port
CMAPI_Network_GetRFInfo()	Get information about the RF channel (frequency, channel, maximum data rate on uplink and downlink)
CMAPI_Network_GetHomeInformation()	Get information about the subscriber's home network
CMAPI_Network_GetServingInformation()	Get information about the serving network the subscriber is connected to
CMAPI_NetConnectSrv_SelectNetwork()	Select a network to connect to
CMAPI_NetConnectSrv_GetNetworkList_Async()	Get the list of available networks
CMAPI_NetConnectSrv_Connect_Async()	Connect to a given network
CMAPI_NetConnectSrv_Disconnect_Async()	Disconnect from a network
CMAPI_NetCon_GetConnectionStatus()	Get the connection status (scanning, connecting, connected, disconnected)
CMAPI_MobileIP_SetState()	Enable or disable mobile IP
CMAPI_DevSrv_GetHardwareVersion()	Get the hardware version of the device
CMAPI_DevSrv_GetFirmwareVersion()	Get the firmware version of the device
CMAPI_DevSrv_GetIMSI()	Get the IMSI of the subscriber
CMAPI_DevSrv_GetMSISDN()	Get the MS-ISDN of the subscriber
CMAPI_DevSrv_GetIMEI()	Get the IMEI of the device

Table 5.8
A Selection of OpenCM API Functions *(Cont.)*

Function	Description
CMAPI_DevSrv_GetRFSwitch()	Get the state of the radio (on/off)
CMAPI_DevSrv_SetRadioState()	Enable or disable the radio (flight mode on/off)
CMAPI_DevSrv_EnablePIN()	Enable or disable the PIN on the SIM
CMAPI_DevSrv_DisablePIN()	
CMAPI_DevSrv_VerifyPIN()	Verify the PIN on the SIM
CMAPI_WLAN_AddKnownNetwork()	Add a known WLAN network
CMAPI_WLAN_Scan_Async()	Scan available WLAN networks and notify application (asynchronous function)
CMAPI_WLAN_Connect()	Connect to a WLAN network
CMAPI_Information_GetSignalStrength()	Get the signal strength
CMAPI_Information_GetIPAddress()	Get the IP address of the device
CMAPI_Information_GetRoamingStatus()	Get the roaming status of the device (home/roaming)
CMAPI_SMS_Send()	Send an SMS
CMAPI_SMS_Get()	Get a received SMS from the device's store
CMAPI_GNSS_SetState()	Switch GNSS receiver on/off
CMAPI_GNSS_GetDevicePosition()	Get the position of the device

Whether or not device manufacturers will allow applications to access all the functionality exposed by the OpenCMAPI remains to be seen. Device operating systems such as Android and iOS® tend to either provide the functions in Table 5.8 as native functions or inhibit them altogether. In neither case will the device manufacturer allow them to be accessible through a third-party API such as OpenCMAPI.

In this chapter we've seen how APIs expose network functionality so that service providers can create value-added applications that rely on IMS, IN, and CAMEL features. On the device side, APIs exist to provide access to device, smart card (SIM), and connectivity features. Standards organizations such as OMA and the 3GPP have published many such APIs. Perhaps not all of these will consolidate as standards, but APIs appear to be here to stay as an approach to opening up networks and terminals to new business models.

References

[1]   Bolton, F., *Pure Corba*, Indianapolis, IN: Sams Publishing, 2001.

[2]   Harold, E. R., and W. S. Means, *XML in a Nutshell*, Cambridge, MA: O'Reilly, 2004.

[3]   Walmsley, P., *Definitive XML Schema*, Upper Saddle River, NJ: Prentice Hall, 2012.

[4]   Thompson, H. S., N. Mendelsohn, D. Beech, and M. Maloney, W3C XML Schema Definition Language (XSD) 1.1 Part 1: Structures, W3C Recommendation http://www.w3.org/TR/xmlschema11-1, 2012.

[5]   Peterson, D., S. Gao, A. Malhotra, C. M. Sperberg-McQueen, and H. S. Thompson, W3C XML Schema Definition Language (XSD) 1.1 Part 2: Datatypes, W3C Recommendation http://www.w3.org/TR/xmlschema11-2, 2012.

[6]   Fielding, R. T., *Architectural Styles and the Design of Network-based Software Architectures*, Ph.D. thesis, University of California at Irvine, 2000.

[7]   Norgaard, J., V. B. Iversen (eds.), "Intelligent Networks and Intelligence in Networks," *Proceedings of the IFIP TC6 WG6.7 International Conference on Intelligent Networks and Intelligence in Networks*, Springer-Verlag, 1996, pp. 296–307.

[8]   Bennett, A., M. Unmehopa, and K. Vemuri, *Parlay/OSA: From Standards to Reality*, Chichester, UK: Wiley, 2006.

[9]   Schulzrinne, H., V. Gurbani, P. Kyzivat, and J. Rosenberg, RPID: Rich Presence Extensions to the Presence Information Data Format (PIDF), IETF RFC 4480, 2006.

[10]   Hardt, D. (ed.), The OAuth 2.0 Authorization Framework, RFC 6749, IETF, 2012.

[11]   Jones, M., and D. Hardt, The OAuth 2.0 Authorization Framework: Bearer Token Usage, RFC6750, IETF, 2012.

[12]   Lodderstedt, T. (ed.), with S. Dronia and M. Scurtescu, OAuth 2.0 Token Revocation, RFC7009, IETF, 2013.

[13]   Open Mobile Alliance, Push, OMA-ERP-Push-V2_3, 2011.

[14]   Brenner, M., and M. Unmehopa, *The Open Mobile Alliance: Delivering Service Enablers for Next-Generation Applications*, Chichester, UK: Wiley, 2008.

[15]   Regulation (EU) No 531/2012 of the European Parliament and of the Council on Roaming on Public Mobile Communications Networks Within the Union, L172, Official Journal of the European Union, 2012.

[16]   Open Mobile Alliance, Guidelines for RESTful Network APIs, white paper, OMA-WP-Guidelines_for_RESTful_Network_APIs, 2014.

[17]   Third Generation Partnership Project, Digital Cellular Telecommunications System (Phase 2+); Specification of the SIM Application Toolkit for the Subscriber Identity Module—Mobile Equipment (SIM—ME) interface, Technical Specification 51.014, 2004.

[18]   Third Generation Partnership Project, Digital Cellular Telecommunications System (Phase 2+); Universal Mobile Telecommunications System (UMTS); LTE; Universal Subscriber Identity Module (USIM) Application Toolkit (USAT), Technical Specification 31.111, 2014.

[19]   Third Generation Partnership Project, Digital Cellular Telecommunications System (Phase 2+); Universal Mobile Telecommunications System (UMTS); Mobile Execution Environment (MExE); Functional description; Stage 2, Technical Specification 23.057, 2012.

[20]   Information Technology—Common Language Infrastructure (CLI), International Standard 23271:2012, ISO/IEC, 2012

[21]   Klyne, G., F. Reynolds, C.Woodrow, H. Ohto, and J. Hjelm, et al., Composite Capability/Preference Profiles (CC/PP): A User Side Framework for Content Negotiation, http://www.w3.org/TR/CCPP-struct-vocab, W3C, 2004.

[22]   Open Mobile Alliance, User Agent Profile, OMA-TS-UAProf-V2_0, 2006.

6

Support Systems

The chapters up to this point have dealt with new and legacy technologies to build digital communications applications and services. But being able to create services and applications is only half the story. A service provider will also have to deploy its services, manage their life cycle, and, most importantly, monetize them.

This chapter provides more insight into the systems that allow a service provider to deploy and manage services, and make a business with them.

6.1 OSS and BSS

The industry generally refers to the systems that manage networks and services as operations support systems (OSSs) and business support systems (BSSs). Although these are distinct terms, they often go hand in hand. OSSs and BSSs have no clear-cut boundary; no definition clearly states where OSSs end and BSSs begin. Many support systems available today combine OSS and BSS functions.

If we have to make a distinction, one could say that OSSs generally manage networks and services, whereas BSSs focus on the customer. Although BSSs and OSSs have no clear-cut boundary between them, they generally offer the following functions when combined:

- *Fulfillment* amounts to capturing and processing customer orders, and provisioning the services that fulfill these orders. Fulfillment involves not only order work flow management (i.e., entering, validating, accepting orders, and financial processing), but also a maintaining service catalog and managing the inventory. The service catalog provides information about services including their pricing, dependency on other services, bundling options, and available service levels. The inventory allows a service provider to keep track of the resources it needs to provide services to its customers such as telephone numbers and SIM cards.
- *Assurance* comprises those actions aimed at making services meet the expected levels of availability, quality, and performance. This involves monitoring services against key performance indicators (KPIs),[1] key quality indicators (KQIs),[2] and service level agreements (SLAs).[3]
- *Billing* allows a service provider to receive money from customers for the use of services.
- *Customer relationship management* (CRM) lets a service provider manage its customer base proactively, with the aim of retaining existing customers, signing up new customers, and making appropriate service offers to each customer.

Figure 6.1 gives a simplified view of the scope of fulfillment, assurance, billing, and CRM. The fulfillment activities transform an order from a customer into a service provisioned to that customer. The assurance functions monitor the service and ensure that it meets the SLA with the customer. The billing function gathers information from the network on the use of the service and creates a bill to send to the customer based on the established rates for the service. CRM concerns the capture and retention of customers.

Taking a simplistic view, one could say that fulfillment and CRM reside more on the BSS side of support systems, while assurance is more an OSS feature. Billing has a role in both OSS and BSS, because it involves keeping

---

[1] Key performance indicators (KPIs) describe parameters that can be measured to determine the performance of an organization or a system. Mean response time, mean time between failure (MTBF), and mean time to recovery (MTTR) are possible KPIs for a system.

[2] Key quality indicators (KQIs) provide a measure for the quality of service offered by a system. KQIs are usually determined by external, subjective observation and cannot be measured within the system. The mean opinion score (MOS) for voice calls represents a typical KQI for a communications system.

[3] A service level agreement (SLA) specifies the level of service provided under a given contract. For example, service levels such as "24 hours a day, 7 days a week support" or "support Mon–Fri, 9AM–5PM," "99% uptime" or "99.99% uptime" may form part of an SLA.

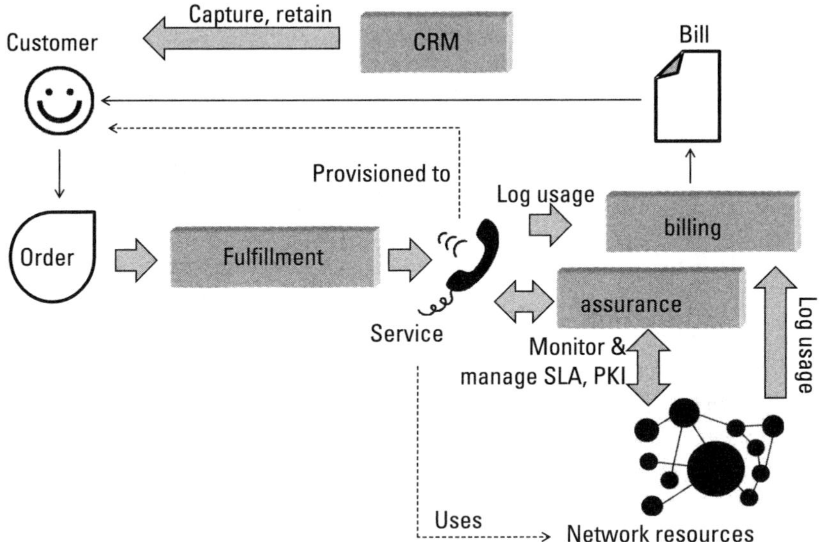

Figure 6.1 Fulfillment, assurance, billing, and CRM.

track of resource usage in the network (OSS) as well as creating bills and sending them to customers (BSS).

Each of these functions—fulfillment, assurance, billing, and CRM—involves complex information models and work flows. As a result, OSS and BSS are large, complex, and expensive systems. In Sections 6.2 and 6.3 of this chapter, we'll have a closer look at the assurance and charging functions, and the standards they use. We won't go into fulfillment and CRM, as they relate more to business aspects, which lie beyond the scope of this book.

6.2 Assurance

Assurance involves at least the following activities:

- *Service activation:* once an order been accepted and processed, the system must configure the service for the customer and register the customer as an authorized user of the service.
- *Quality management:* monitoring and (re)configuring service and network components to ensure that services meet the SLAs with the customer and the associated KPI and KQI. KPI and KQI differ in that KPI provides internal measures while KQI provides external measures

of success. For example, KPI may include mean time between failure (MTBF) and mean time to recovery (MTTR). KQI, on the other hand, may involve customer rating of call quality. Whereas KPI and KQI generally apply to services and networks, SLAs apply to customers. An SLA may specify, for example, that a customer gets a discount if the mobile network causes a call to drop involuntarily or if the data rate drops below an agreed minimum.

- *Problem management:* defines the work flow to deal with problems at network or service level. Customer complaints as well as network-generated alarms can trigger problem management activities, which in turn can involve human or automated intervention. In most cases, problem management involves a trouble ticketing system to keep track of problems and their resolution.

Assurance requires significant infrastructure to monitor service and network resources and to manage their configurations and performance. The assurance activities rely on network management techniques, which we'll discuss in the sections that follow.

6.2.1 Network Management

As long as telecommunications networks exist, there has been a need to manage them. Standards organizations began publishing standard models and protocols for network management in the early 1990s, as digital networks proliferated and the need for network management standards increased. Two standards stand out:

- The OSI operation, administration, and management (OAM) model with fault, configuration, accounting, performance, and security (FCAPS) management as specified in the ISO 10164 and 10165 series of standards;
- The ITU-T telecommunications management network (TMN) model as specified in the ITU-T M.3000 series of recommendations.

Although these are different models with different protocols, they have essential concepts in common and have been converging over the years as a result of joint efforts by ISO and ITU-T. In this section we'll focus on their similarities rather than their differences.

ITU-T recognizes four management layers:

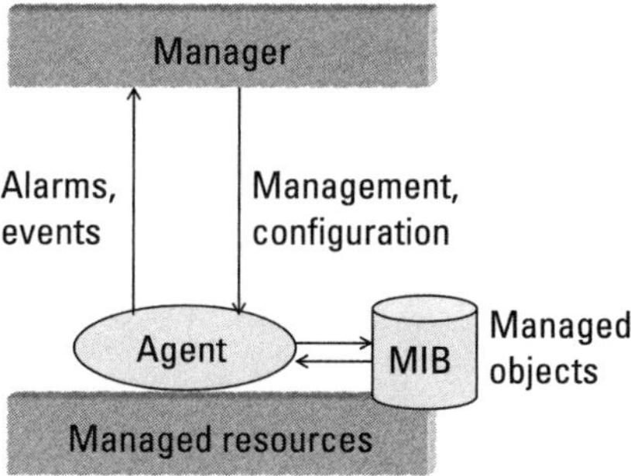

Figure 6.2 The agent-manager model.

- *Element management* involves monitoring and configuration of individual network elements such as routers, switches, and telephone exchanges.
- *Network management* deals with supervision and configuration at a more global network level.
- *Service management* describes the activities to configure, activate, and provision services, and to ensure their quality.
- *Business management* deals with business processes such as service inventory, order management, and billing.

ITU-T and ISO have so far produced mature standards[4] only for element management, and to a lesser degree for network management. The TM Forum has more recently begun to address the higher management layers in its Frameworx® suite of specifications. In the following sections, we'll take a closer look at the standards that have emerged in these areas.

6.2.2 The Agent-Manager Model

Standards and commercially available systems for element management have existed since the early 1990s, and still make up for the majority of network management systems in operation today. Both ITU-T and ISO follow an agent-manager model. Agents represent the management software that interacts with a network resource to monitor and configure it. Managers represent

---

[4] Mature standards in the sense that they actually specify protocols and information models beyond some general concepts.

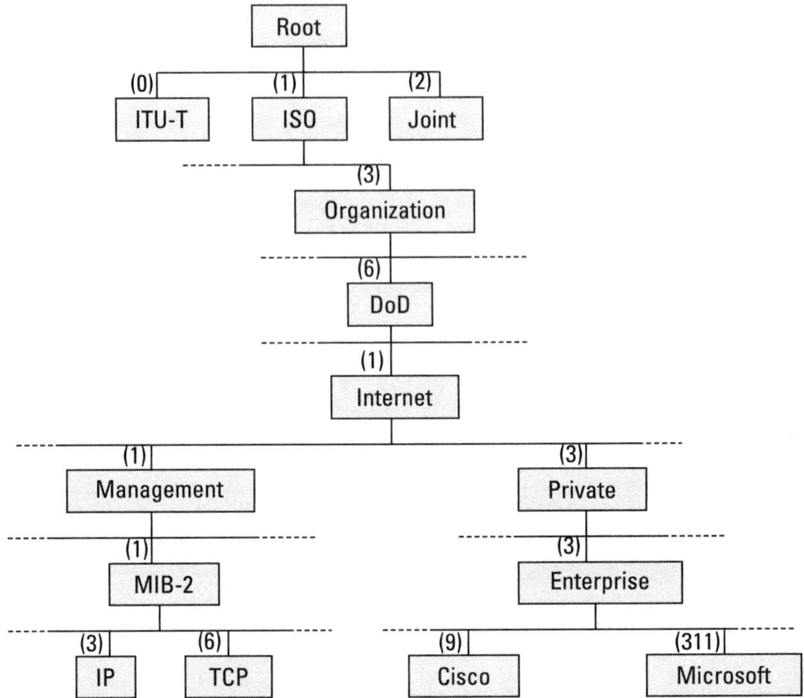

Figure 6.3   MIB tree as defined by ISO and ITU-T

the management applications that interact with agents. Figure 6.2 illustrates the agent-manager model.

Both ITU-T and OSI have an object-oriented approach to element management whereby managed objects (MOs) represent the managed network resources. A manager can query and configure network resources through their MOs.

Agents access MOs through a management information base (MIB), a tree structured repository that stores the MOs. Figure 6.2 also illustrates the role of the MIB in the agent-manager model.

Each MO has a unique object identifier derived from its path in the MIB. MIBs have a standardized structure defined by ITU-T, ISO, and other organizations such as the IETF. Figure 6.3 shows a fragment of the MIB tree defined by these standard organizations.

The management subtree on the left in Figure 6.3 contains MOs for standard internet protocols such as IP, TCP, UDP, and ICMP. The enterprise subtree on the right contains MOs for proprietary network equipment from different vendors. Following the MIB structure in Figure 6.3, a MO to manage an IP stack deployed in a router, host, or other equipment has the following object identifier:

```
.iso.organization.dod.internet.management.mib-2.ip
```

6.2.3 Managed Objects

A MO has variables that a manager can read or write. Table 6.1 shows some of the variables of the MO for an IP stack (this is a sample, and the list is not exhaustive).

Table 6.1
Some Variables of the MO for an IP Stack

Variable	Read/Write	Description
ipDefaultTTL	Read/write	Allows a manager to get and set the default time to live (TTL)[1] for IP packets
ipInReceives	Read only	The number of packets received
inInAddrErrors	Read only	The number of packets received with an address error (i.e., this IP stack should not have received this packet based on its address information)
ipInDiscards	Read only	The number of packets discarded when received (e.g., due to overload)
ipForwDatagrams	Read only	The number of packets forwarded (forwarding is what routers do, as we saw in Section 3.1.1)
ipReasmFails	Read only	The number of packets for which reassembly has failed[2]

[1] The header of an IP packet has a one-octet field called time to live (TTL) that counts number of hops the packet has made between routers. When a host sends an IP packet, it sets the TTL to a value between 0 and 255. Each router on the packet's path decreases the counter by one. A router that finds the counter at 0 has to discard the packet. This helps to avoid congestion as the result of (unintended) forwarding loops between routers, which could cause packets to go around forever. On the downside, it can also lead to packet loss if a packet's route has more hops than its TTL allows.

[2] Different physical networks can put different limitations on the maximum size of a packet. This can cause problems if, for example, a packet originates in a network that can handle packets of 1,500 octets and has to pass through a network that can only handle packets of 512 octets. To solve this problem, IPv4 allows a router to fragment a packet into several smaller ones that fit the network they're forwarded to. As a consequence, the host at the receiving end will have to reassemble fragmented packets to obtain the original packets sent. Note that IPv6 does not support fragmentation and simply relies on the originating host to retransmit packets with a smaller size if routers on the path complain about size. This causes more packet loss and more retransmission, but makes IPv6 routers simpler and faster.

ISO and ITU-T use the abstract syntax notation 1 (ASN.1)[5] to describe MOs. ISO and ITU-T standardized ASN.1 in the 1980s as a formal language to specify structured data and have used it mostly to specify the format of protocol messages. The following is a simple example of a MO description in ASN.1 for a MO that represents a firewall from MyCompany:

```
MYFIREWALL-MIB DEFINITIONS ::= BEGIN

IMPORTS
        enterprises
                FROM RFC1155-SMI
        OBJECT-TYPE
                FROM RFC-1212
        DisplayString
                FROM RFC-1213;

mycompany OBJECT IDENTIFIER ::= {enterprises 1234}
firewall OBJECT IDENTIFIER ::= {mycompany 3}

firewallControl OBJECT-TYPE
  SYNTAX   INTEGER  {active (1), inactive (2)}
  ACCESS   read-write
  STATUS   mandatory
  DESCRIPTION
    "This variable controls the state of the
firewall. To begin filtering set it to active (1).
To stop filtering, set it to inactive (2)."
  ::= {firewall 1}

firewallState OBJECT-TYPE
  SYNTAX   INTEGER {
                    no-attack-detected (1),
```

```
                          port-scan-detected (2),
                          ping-attack-detected (3),
                          syn-attack-detected (4),
                          ip-session-attack-detected (5),
                          firewall-error (6)
                        }
      ACCESS    read-write
      STATUS    mandatory
      DESCRIPTION
         "This variable describes the state of the
   firewall, whether it is detecting any type of
   attack, or whether it is in error state."
      ::= {firewall 2}
   END
```

MOs for real equipment usually have much more complex ASN.1 descriptions that require hundreds or even thousands of lines. Many vendors provide their own proprietary MOs that do not (fully) follow the ISO and ITU-T standards.

6.2.4 Network Management Protocols

ISO and ITU-T jointly specified the common management information protocol (CMIP) [1, 2] to manage network resources through their MOs. CMIP forms part of the ISO open systems interconnect (OSI) protocol stack. Although ISO didn't design it for use on IP networks, CMIP is now commonly used with TCP/IP.

In the late 1980s, IETF specified a simpler protocol for the management of IP network equipment called the simple network management protocol (SNMP) [3]. SNMP and CMIP differ in that SNMP provides mostly monitoring functions, while CMIP allow a manager to perform more comprehensive management operations on an MO. CMIP also provides reliable transport and better security than SNMP. But in spite of these differences, SNMP is much more widely used thanks to its simplicity and its tight fit with IP protocols. Table 6.2 compares the CMIP and SNMP primitives on a high level. It shows that CMIP supports operations to create and delete MOs, while SNMP does not have matching operations. CMIP also provides an operation to provide management actions on MOs, which does not have its counterpart in SNMP.

Table 6.2
A High-Level Comparison of SNMP and CMIP Primitives

CMIP	SNMP	Description
mCreate	—	Create a new MO
mDelete	—	Delete a MO
mGet	getRequest	Get the variable(s) of an MO
mSet	setRequest	Set the value of a variable or variables of an MO
mAction	—	Perform a management action on an MO (the protocol message specifies the action)
mEventReport	trap	Report an event (from agent to manager)

6.2.5 Frameworx

While the network-related functions of OSS and BSS draw on network management technologies and standards, few standards exist for service management and business support. The TM Forum has begun documenting best practices for business flows in telecommunications, service, and customer management, called Frameworx. Frameworx consists of a loosely couple of specifications and best practice documents that can be used on their own or in combination. Some of the key components of Frameworx are as follows:

- The *enhanced telecommunications operations map* (eTOM) describes a layered view of business processes for running a telecommunications enterprise, including high-level processes for marketing, service development, resource development, and supply chain management.
- The *shared information and data* (SID) framework defines an information model that helps service providers and equipment vendors establish common terminology and definitions for all things that have to be managed.
- The *multitechnology operations system interface* (MTOSI) specifies an XML format for the exchange of data between the different components of an OSS and BSS.
- *OSS/J* specifies a Java interface between distributed OSS and BSS components. Its purpose is similar to MTOSI, but it has different messages and it's bound to Java.

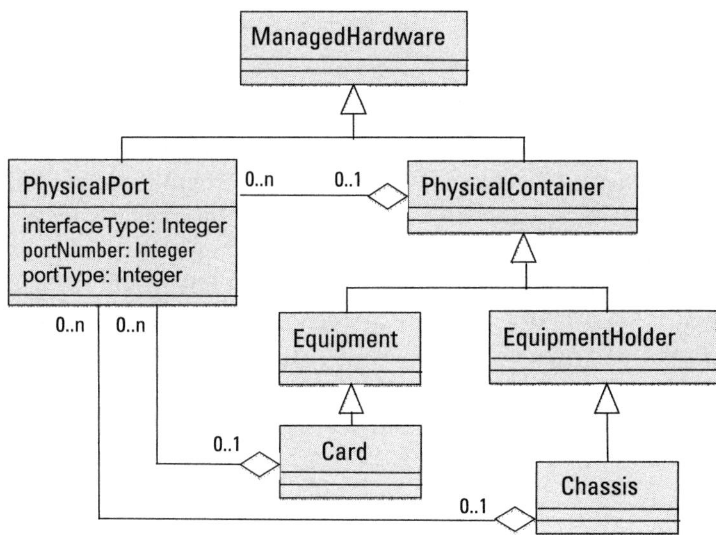

Figure 6.4 Example SID information model.

SID [4] describes communications business and network-related objects and their relations. Figure 6.4 shows the example of a SID model for a (hardware) port consisting of line cards mounted in a chassis.

Figure 6.4 uses UML[6] notation, whereby the boxes represent classes, triangles represent inheritance, and diamonds represent composition. The diagram in Figure 6.4 specifies that a chassis is a kind of equipment holder, which in turn has the properties of a physical container. A physical port consists of zero or more chassis and holds zero or more line cards.

---

[6] The unified modeling language (UML) provides a collection of diagrams to support object-oriented design and programming. The UML diagrams fall in two categories: structure diagrams and behavior diagrams. Structure diagrams describe the static aspects of an object-oriented system, such as class inheritance. Behavior diagrams describe dynamic aspects such as the message flow between objects. The diagrams are mostly graphical, but also have a textual (XML) representation so they can be imported, exported, and exchanged between tools. Although UML offers more than 15 diagrams, probably the most popularly used ones are the use case diagram, class diagram, and sequence diagram. Contrary to popular belief, UML does not represent any methodology for creating object-oriented software; it merely provides a collection of diagrams. However, there are standard and proprietary software development methodologies that build on UML diagrams. UML emerged in the mid-1990s as the fusion of the object-oriented design theories of Grady Booch, James Rumbaugh, and Ivar Jacobsson and is now an OMG standard.

6.2.6   Multitechnology Operations System Interface

Until the late 1990s, most operators and service providers deployed large-scale, monolithic BSS and OSS, usually from a single vendor. Advances in distributed computing (in particular WS) and increasing competition in the telecommunications market caused these monolithic OSS and BSS to give way to distributed systems with loosely coupled components from different vendors. With MTOSI, the TM Forum sought to standardize the interface between distributed OSS and BSS components. Although MTSOSI standardizes the message formats, it does allow vendors to extend the messages with proprietary data structures. Figure 6.5 illustrates how MTOSI interconnects BSS and OSS components.

MTSOSI uses a client-server model whereby one OSS component makes a request to another and the other responds, and uses SOAP messages to transport the requests and responses. The following example shows an MTOSI request to an inventory management system to provide the names of all equipment in the inventory:

```
<Envelope>
  <Header>
    <MTOSIHDR tmf854Version="1.0">
      <domain>Inventory</domain>
      <activityName>getInventory</activityName>
      <msgName>getAllEquipmentNames</msgName>
      <msgType>REQUEST</msgType>
      <payloadVersion>1.0</payloadVersion>
      <senderURI>
      http://myClienOSSComponent.net
      </senderURI>
```

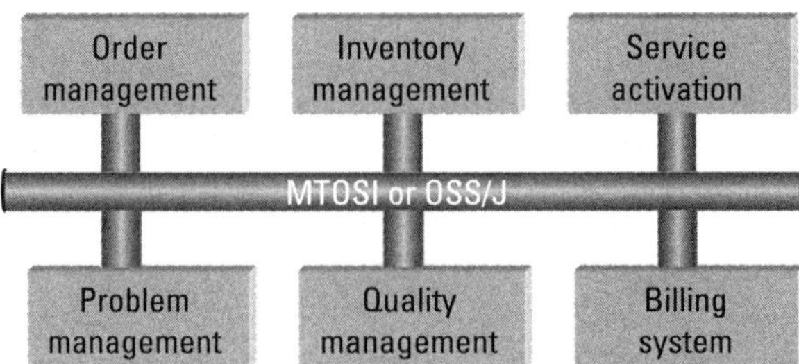

Figure 6.5   Using MTOSI to interconnect BSS and OSS components.

```
        <destinationURI>
          http:// theServingOSSComponent.net
        </destinationURI>
        <correlationId>0001</correlationId>
        <communicationPattern>
          MultipleBatchResponse
        </communicationPattern>
        <communicationStyle>
        MSG
        </communicationStyle>
        <requestedBatchSize>0</requestedBatchSize>
      </MTOSIHDR >
    </Header>
    <Body>
      <getAllEquipmentNames tmf854Version="1.0" />
    </Body>
  </ Envelope>
```

The example shows how MTOSI makes extensive use of SOAP headers to provide meta-information about the request, such as the URI of sender and receiver and the type of response solicited. Most web services today don't tend to use the optional SOAP header, and the MTOSI specifications follow an atypical (but technically correct) approach in this respect.

MTOSI and OSS/J can work on different levels of granularity. Figure 6.5 shows how MTOSI interconnects high-level OSS components, but it can also interconnect different modules within an inventory management or billing system, for example.

6.2.7 Service Delivery Platforms

A service delivery platform (SDP) provides a hardware and software environment to create, deploy, run, and manage services. SDP [5] does not have a unique definition in industry and does not have standardized interfaces or components. All major telecommunications vendors have a SDP in their offering, but each vendor defines and implements it in its own way.

Depending on how one defines SDP, one may consider it a part of the service assurance functions or a plain service execution platform. In this section we'll take the first approach and treat it as a component that contributes to service assurance.

The industry started using the term SDP toward the turn of the millennium, as IN started giving way to OSA APIs and IMS. The IN architecture has a well-defined service creation and deployment model, as we saw in Chapter 2. The IN standard describes how SIBs combine to build services on the GFP, and how the FEs host service components and trigger services at the DFP. The IN standard also describes the service creation environment (SCE) as the software development environment where the service designer composes SIBs into services.

By contrast, the 3GPP OSA/Parlay specifications describe only APIs, not a model for the design or deployment of applications that use them. Nor do the IMS specifications define how a service provider can program a SIP application or deploy it on an AS. The industry filled this gap by coming up with the concept of the SDP.

Looking at the way vendors define their SDPs, we can distinguish two approaches. One is a narrower view whereby the SDP consists of the tool chain to implement services and applications in a software design environment, and then deploy and run them in a run time environment. This type of SDP usually interfaces with the functional entities in the network through OSA, OMA, OneAPI APIs, or even raw SIP.

Other vendors take a broader view and consider the SDP to consist of the software development and run time environment, plus the OSS, BSS, and billing system. Figure 6.6 illustrates the concept of the SDP in both the broader and narrower sense.

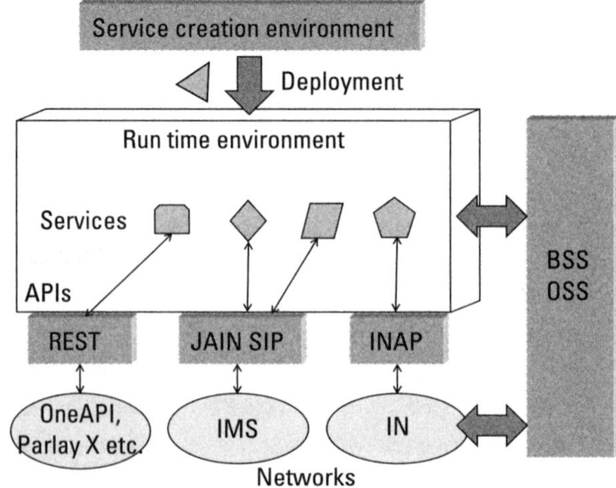

Figure 6.6   Service delivery platform.

If you've read the previous chapters, it may strike you that Figure 6.6 looks a lot like Figure 4.15 (JAIN SLEE) and also has a lot in common with Figures 4.11 (SIP servlets) and 5.6 (network APIs). This is not a coincidence. SDPs provide the execution environment for many of the technologies we've seen so far, including JAIN SLEE, SIP servlets, JAIN SIP, and even the IN and CAMEL SCP. In fact, JAIN SLEE comes close to being a de facto SDP standard.

In the 1990s, many vendors tried to create graphical SCEs that would allow service designers to create new services by plugging together predefined graphical components on a screen. The IN standards promoted this way of service creation by picturing SIBs as components with wel-defined inputs and outputs that can be chained together to build services.

In practice this proved feasible only for very simple services (e.g., those that use simple "if-then" rules to translate the destination number for a call). Services of even moderate complexity need extensive configuration and exception handling that diminishes the effectiveness of graphical composition. Some services from the IN era required on the order of 1,000 SIBs to cope with all possible user interactions, flows, and exceptions.

Service creation involves programming, regardless of whether supported by graphical tools. Graphical tools can help, but they can't do magic. This means that service creation for IN, CAMEL, and IMS still remains very much a specialist task—not something anybody can do by clicking and connecting icons on a screen.

6.3 Charging and Billing

No doubt one of the most important requirements for a service provider is to be able to charge for services used and bill subscribers for them. Charging and billing form part of the OSS and BSS, but they are of such importance that we'll dedicate a section to it. Charging and billing represent the pillar on which the business of a service provider rests.

Many people in the industry use the terms *charging* and *billing* interchangeably, but they have a slightly different meaning. Following their dictionary definitions, charging is making a note of a sum of money as being owed; billing is the act of submitting to a debtor an account of money owed.

In communications systems, charging means keeping track of the use of network resources by a subscriber and the cost associated with it. Billing consists of documenting the cost of the service in a bill and presenting it to the subscriber for payment. By consequence, one can argue that charging functions reside at network management level (OSS), while billing functions lie more on the business side (BSS).

6.3.1   CDR and IPDR

To be able to bill a subscriber for services, a service provider must know details about the resources the subscriber has used. To provide this information to the service provider, network equipment create charging data records (CDRs)[7] for the resources used. Both 3GPP and ITU-T have standardized CDR formats, where the 3GPP specifications focus on cellular networks and IMS.

A CDR is made up of attribute-value pairs[8] that identify the subscriber, the resources, or services he or she has used; the duration of use; and other details relevant for billing. Each service requires different attributes for charging, so CDRs have different formats depending on the service. For example, a voice call requires recording of the call duration, while a short message (SMS) requires recording of the discrete time the message was sent.

Table 6.3 shows a sample of the mandatory CDR fields as specified by 3GPP for a mobile originated call. The CDR can have many additional optional fields such as the radio channel used and CAMEL services invoked, omitted here for simplicity.

The 3GPP has specifications (TS 32.297, 32.298) for the general format of CDRs that also describe how to group CDRs in files and transfer them to a charging database. The 3GPP defines CDRs in abstract terms, by listing the fields and their datatype, and allows for several encodings including binary formats and plain text formats such as XML.

The 3GPP does not have a single standard that specifies CDRs. Instead, each service standardized by the 3GPP has a charging-related specification that describes the specific format of CDRs for that service. Table 6.4 gives an overview of some of the general charging specifications by the 3GPP and some of the charging specifications that are specific to services. Table 6.4 does not provide an exhaustive list of all 3GPP charging specifications, but it lists the most important ones.

The TM Forum has specified similar format to record chargeable use of resources in IP networks, called *IP detail records* (IPDRs). IPDRs allow a service provider to record chargeable events related to VoIP calls, video streaming,

---

[7] In the era of switched telephony, CDR meant call detail record. The 3GPP changed this to charging data record to adapt the concept to IMS networks, where not all sessions are (voice) calls.

[8] The 3GPP calls attributes "fields," hence the left column in Table 6.3 has the header "Field."

Table 6.3
CDR Fields for a Mobile Originated Call (3GPP)

Field	Description
Record Type	Type of CDR, in this case "mobile originated"
Served IMSI	IMSI of the calling party
Served IMEI	IMEI of the device of the calling party
Served MSISDN	MS-ISDN of the calling party
Called number	The address of the called party
Recording Entity	Address of the MSC that handled the call and created this CDR
Location	Identifier of the cell the calling party was connected to
Basic service	The bearer service used to connect calling and called parties
Classmark	The class mark[1] of the device of the calling party
Call duration	The duration of the call (in case the call was successful), or the holding time (in case the call could not be completed)
Cause for termination	Indicates whether the called party, calling party, or a network event terminated the call
Call reference	A unique identifier, to distinguish different CDRs for the same calling party

[1] The class mark is a standard binary format for information about the capabilities of a mobile device, such as the data capabilities (2.5G, 3G, LTE), frequency bands, encryption algorithms, positioning methods, and character sets it supports.

web services, and other services offered over IP networks. IPDRs consist of attribute-value pairs and can have various encodings, just like CDRs.

The following is an example of an XML-encoded IPDR that records the number of octets sent (89096) and received (12789) by subscriber Zuidweg032783789 on a device with IP address 192.168.3.231. The nasIdentifier field identifies the equipment that generated the IPDR by its IP address:

```
<IPDR>
   <subscriberId>Zuidweg032783789</subscriberId>
   <ipAddress>192.168.3.231</ipAddress>
   <nasIdentifier>192.168.0.1</nasIdentifier>
   <acctInputOctets>12789</acctInputOctets>
   <acctOutputOctets>89096</acctOutputOctets>
</IPDR>
```

Table 6.4
3GPP Charging Specifications

3GPP Specification	Number
General charging specifications	
Charging architecture and principles	32.240
CDR file format and transfer	32.297
CDR parameter description	32.298
Diameter[1] charging application	32.299
Charging specifications for specific services	
Circuit switched domain charging	32.250
Packet switched domain charging	32.251
IMS charging	32.260
MMS charging	32.270
Location services charging	32.271
SMS charging	32.274
MMTel charging	32.275

[1] For a brief explanation of the Diameter protocol, see Section 6.3.3.

6.3.2 Mediation and Rating

A network tends to generate an overwhelming amount of CDRs or IPDRs. Even a single service such as a voice call between two parties generates a torrent of records.

Consider the example of an IMS voice call between two roaming mobile subscribers shown in Figure 3.14 (Section 3.4.2). Such a voice call will generate CDRs from the following network elements:

- The P-CSCFs in both the calling and called party's visited networks;
- The S-CSCFs in both the calling and called party's home networks;
- The ASs in both the calling and called party's home networks (if there is any additional service processing);
- The I-CSCF in the called party's home network;
- The HSS in the called party's home network;
- The SGSNs in both the calling and called party's visited networks that transport the data.

And that is only for a single call. A (mobile) network can process millions of calls each day, depending on its size. Without exaggeration, one can consider CDRs big data, which makes charging and billing a data- and processing-intensive exercise. In the not-so-distant past, charging and billing systems required expensive mainframe computers. Only in recent years have low-cost distributed computing and cloud computing brought their cost down.

The biggest challenge lies in correlating the CDRs generated for a service in different parts of the network. This requires analysis and correlation of individual fields of the CDRs so as to determine to which subscriber and service they belong and what resource use they record. This task is called *mediation*, and it's effectively big data processing.

In order to bill a subscriber, the service provider must put a price on the services consumed and their use of network resources. A rating engine takes care of this task. Figure 6.7 illustrates the role of mediation and rating in the billing process.

Figure 6.7 shows how several network elements such as the SGSN and S-CSCF all generate CDRs for a single service session. If the service involves streaming from a content server, this server may generate IPDRs, too. The mediation function correlates all these CDRs and IPDRs, and the rating engine determines the price to bill to the subscriber.

The rating engine establishes prices for several aspects of a service:

- *Time:* telephone calls are typically charged by time (and destination);
- *Volume:* mobile data traffic is usually charged by volume;
- *Events:* such as sending a short message (SMS).

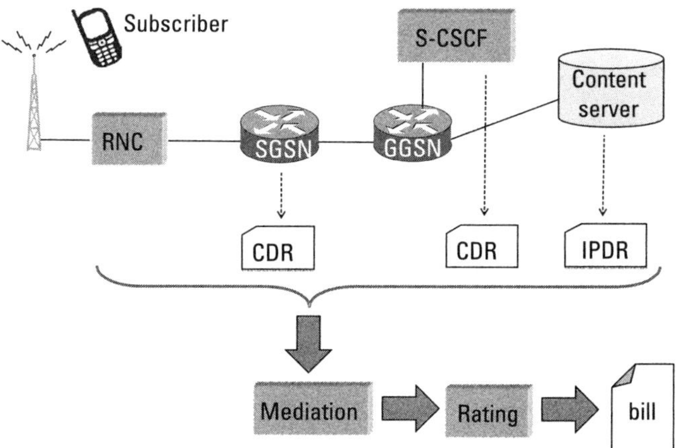

Figure 6.7 CDR collection, mediation, rating, and billing.

The rating engine also calculates cross-service discounts. For example, it may apply a $0 charge for calls to the service providers' help desk, or apply a lower rate per minute for voice calls if a subscriber also has a data plan.

6.3.3 Charging in IMS

As Table 6.4 illustrates, the 3GPP has an extensive suite of specifications that cover charging for both circuit switched and packet switched networks. The 3GPP distinguishes between two types of charging:

- *Offline charging:* the billing system processes the CDRs after use of the service;
- *Online charging:* the billing system processes CDRs in real time and can adapt mediation and rating while the service is in use.

Offline charging represents the classic charging model as applied in telephone networks for decades. It remains the dominant charging model for packet switched networks and IMS services, also. Figure 6.8 shows the 3GPP architecture for offline charging.

As shown in Figure 6.8, the 3GPP architecture for offline charging has three main functions:

- *Charging trigger function* (CTF) associated with each network element generates charging events and sends them to the CDF.
- *Charging data function* (CDF) creates CDRs from the charging events received from the CTF.

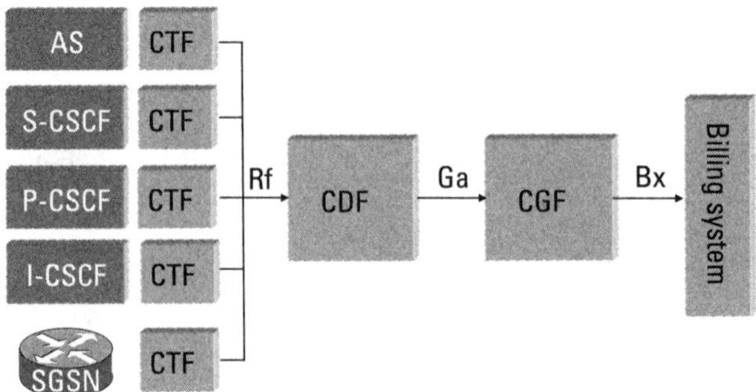

Figure 6.8 The 3GPP offline charging architecture.

- *Charging gateway function* (CGF) transfers CDRs from the CDF to the billing system for processing.

The 3GPP defines the CTF, CDF, and CGF as functions, not physical nodes. A service provider can collocate or distribute these functions as needed. For example, a network element that generates CDRs can host all three of the CTFs, CDFs, and CGFs and connect directly to the billing system. An alternative configuration could collocate the CDFs and CGFs on a node dedicated to charging, connecting to the network elements' CTFs on one side and to the billing system on the other.

R_f, G_a, and B_x are the names of the interfaces between the CTF and CDF, CDF and CGF, and between the CGF and billing system, respectively. IMS uses the Diameter protocol on the R_f interface to send charging events to the CDF.

Diameter is a protocol specified by IETF [6] that provides a general purpose message structure for authentication, authorization, and accounting (AAA) procedures. Diameter messages have a binary format and consist of a command code and a list of attribute-value pairs. IANA[9] has registered attributes for a broad range of AAA procedures and allows organizations to register new attributes as needed for new applications of the protocol. Diameter messages for IMS charging mostly use the `Accounting-Request` and `Accounting-Response` command codes.

6.3.4 Correlating IMS Charging Data

Each IMS element, including the S-CSCF, P-CSCF, I-CSCF, and AS, generates CDRs with different content as specified by 3GPP TS 32.260 [7]. One of the main challenges, as we saw in Section 6.3.2, is to correlate all the CDRs for an IMS session, generated by the IMS elements involved and the underlying packet network.

To enable correlation and mediation, the CDRs for IMS contain the following data:

- *IMS charging identifier* (ICID) provides a unique identifier for the session and user. The first IMS functional entity that intervenes in the session (usually the P-CSCF) generates this identifier and passes it on to subsequent IMS functional entities on the session's signaling path.

---

[9] The Internet assigned numbers authority (IANA) is the Internet Society's administrator for resources for the Internet protocol, such as IP addresses, TCP ports, and Diameter attribute codes. IANA is a sister organization of the IETF, as both belong to the Internet Society.

- *Access network charging identifier* provides information on the underlying packet network elements involved in the session, including the SGSN or GGSN address and the PDP context parameters.
- *Interoperator identifiers* (IOI) identify the originating and terminating networks involved in the service. If the service requires traffic to pass through transit networks, these are also identified.

The IMS elements involved in a session pass these correlation identifiers in the `P-Charging-Vector` SIP header. This way the CDRs generated by each of the IMS elements on the signaling path can be correlated. The following example shows a `P-Charging-Vector` inserted into a SIP `INVITE` message forward by a proxy such as a S-CSCF. The other SIP `INVITE` header details are omitted for simplicity:

```
INVITE … SIP/2.0
Via:…
From:…
To:…
Call-ID:…
CSeq:…
P-Charging-Vector: icid-value=d3498y38ybe4b3946;
                   icid-generated-at=192.168.0.10;
                   orig-ioi=myhomenetwork.net
```

In this example, the IMS entity with IP address `192.168.0.10` that initiated the signaling (possibly the S-CSCF in the subscriber's home network) created the ICID `d3498y38ybe4b3946` (just a unique identifier) and inserted it into the SIP `P-Charging-Vector` header, together with the URL `myhomenetwork.net` identifying the network where the session originated. Any transit networks and the network where the session terminates will add their `orig-ioi` to the `P-Charging-Vector`.

6.3.5 Online Charging

Online charging allows for more dynamic billing models that stimulate competition between service providers, as they can adapt their rates in real time. It also enables prepaid services,[10] as it allows the charging system to interact with a service and terminate or disable it when a subscriber runs out of credit.

---

[10] Sections 2.9.1, 5.3.4, 5.4.3, and 5.5.2 show how CAMEL and network APIs can be used to create prepaid calling services. Online charging provides an alternative enabler for prepaid calling services.

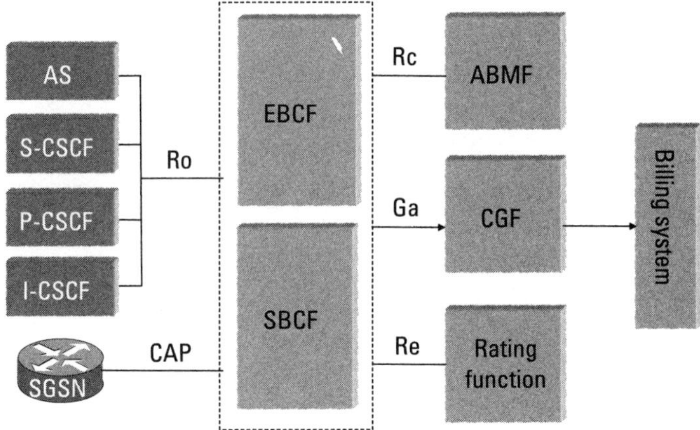

Figure 6.9 The 3GPP online charging architecture.

Online charging requires high-performance hardware and software, which has become more affordable as information technologies advance. Figure 6.9 shows the 3GPP architecture for online charging.

The 3GPP architecture for online charging defines functions as for offline charging, but the model differs. The online charging model distinguishes between event-based charging and session-based charging functions (EBCF and SBCF), which have a more interactive role than the SDF in the offline charging architecture.

The EBCF handles discrete charging events such as outgoing call attempts, short message delivery, and roaming requests (when a subscriber powers his or her device on in a visited network). The SBCF handles session-related charging such as charging for voice or video calls.

Both the EBCF and SBCF can interact with network elements that make charging requests, and they can instruct the network to deny a service when a subscriber has insufficient credit. This marks a difference from the offline charging model, where the CDF only collects chargeable events to create CDRs for offline processing. The SBCF can monitor an ongoing session and request its termination if a subscriber runs out of (prepaid) credit.

The EBCF and SBCF interact with the account balance management function (ABMF) to check a subscriber's credit, to reserve funds on an account, and to charge them. Online charging provides an alternative to the CAMEL or OneAPI implementations of prepaid calling discussed in Sections 2.9.1 and 5.5.2.

To be able to charge in real time against subscriber accounts, the EBCF and SBCF connect to the rating function to obtain rates for the services requested. This real-time interaction with the rating function allows a service provider to adapt the rates dynamically.

The EBCF and SBCF also generate CDRs for post-service processing and offline billing. This aspect of the online billing architecture is the same as for offline billing.

References

[1]   Information technology—Open Systems Interconnection—Common Management Information Protocol: Specification, Recommendation X.711, ITU-T, 1998.

[2]   Information technology—Open Systems Interconnection—Common Management Information Protocol: Specification, International Standard 9596-1, ISO/IEC, 1998.

[3]   Case, J., M. Fedor, M. Schoffstall, and J. Davin, A Simple Network Management Protocol (SNMP), RFC 1157, IETF, 1990.

[4]   Reilly, J. P., *Implementing the TM Forum Information Framework (SID)*, Morristown, NJ: TM Forum, 2012.

[5]   Ahson, S. A, and. M. Ilyas (eds.), *Service Delivery Platforms*, Boca Raton, FL: Auerbach Publications, 2011.

[6]   Fajardo, V., J. Arkko, J. Loughney, and G. Zorn, Diameter Base Protocol, RFC 6733, IETF, 2012.

[7]   Third Generation Partnership Project, Digital Cellular Telecommunications System (Phase 2+); Universal Mobile Telecommunications System (UMTS); LTE; Telecommunication Management; Charging Management; IP Multimedia Subsystem (IMS) Charging, TS 32.260, 2014.

7

Things to Come

Telecommunications are changing, not only in terms of technology but also in terms of business. The changes that took place over the past two or three decades have been difficult to foresee, and analysts have made several false predictions over time.

Twenty-five years ago, nobody thought that the Internet would be widely used for telephony. On the other hand, many futurists thought that video calling would be a standard service provided by network operators in the 1990s. But video calling didn't really gain widespread acceptance until after 2005 and now mainly exists as a free service over the Internet rather than a paid-for operator service with guaranteed QoS.

Fifteen years ago, big telecommunications equipment manufacturers such as Alcatel-Lucent believed that ATM would eventually replace IP, because ATM provides faster switching and better control of QoS, which made it much more suitable for voice telephony, video conferencing, multicast streaming, and other real-time and quality-critical services. But the market remained loyal to IP because of its significantly lower deployment cost. And who saw Apple's rise from the ashes and the smart phone revolution coming at the turn of the millennium?

7.1   Predictions from the Turn of the Millennium

Making predictions for the future, no matter how much they draw on trends that seem unavoidable, is risky business. In the last chapter of his 2002 publication *Next Generation Intelligent Networks* [1], the author of this book identified some of the trends and predictions of that time. Table 7.1 summarizes them and appraises to what degree they became reality.

At first glance, Table 7.1 appears to show that the industry made fairly accurate predictions at the turn of the millennium of the direction things would take in the following fifteen years. Network convergence, all-IP networks, distributed applications, and APIs consolidated over the years. But the predictions that did not come true and trends that were overlooked proved to have disruptive impact on the industry. The emergence of smart phones and OTT applications stands out as one of the most disruptive developments of the last decade. The failure of TINA, in which many operators and manufacturers invested millions of dollars throughout the 1990s, and the reluctant deployment of IMS also show how the industry diverged from the predicted path.

At the risk of making inaccurate predictions, we'll dedicate this chapter to the discussion of things to come in telecommunications.

Table 7.1
Accuracy of Predictions Made About 15 Years Ago

Prediction	Accuracy
Convergence of fixed and mobile networks and Internet.	Yes.
Evolution toward an all-IP network.	Yes.
Implementation of services with distributed object-oriented middleware (CORBA).	Partly. Distributed services and applications are now mainstream but use WS rather than CORBA.
Distribution of service control over the network and devices.	Yes, but OTT applications rather than operator provided services (such as IN, CAMEL, and MExE) are now driving the shift of intelligence toward devices and the cloud.
APIs.	Yes, although deployment remains slow and reluctant.
Adoption of Telecommunications Information Networking Infrastructure (TINA) principles.	No. The less disruptive, less abstract, and more IP oriented IMS specifications became the standard.

7.2 Machine-to-Machine Communications

Mobile networks made a source of steady profit in the 1990s and early 2000s, and had spectacular subscriber growth. In present times, however, mobile devices have a penetration of 100% worldwide and well over 100% in many countries, leaving little margin for growth. Meanwhile, increasing competition and regulation puts pressure on prices, making it difficult for network operators to increase revenues.

As human subscribers leave little more room for expansion, operators have set their eyes on machine-to-machine (M2M) communications for new growth opportunities.

7.2.1 M2M Opportunities and Challenges

The world has billions of devices to connect, from cars to security systems and from traffic lights to medical devices. M2M communications have applications in areas such as

- *M-payment:* electronic points of sale (POS) are among the most established M2M applications, but innovative new M-payment applications emerge such as mobile payment at vending machines.
- *Automotive:* connected cars that interact with their owners, process online traffic data, and can communicate maintenance data directly to the garage.
- *E-health:* medical sensing devices used at home and in the field that remotely send medical data to doctors and hospitals. In the future, doctors may even be able intervene remotely through such devices.
- *Power grids:* smart meters and remote sensors and actuators that facilitate power grid management.
- *Smart buildings:* remote sensors and actuators that improve the energy efficiency of buildings and enable security and smart home applications.
- *Smart cities:* remote control of street lighting, traffic lights, sensing air quality, noise levels, traffic and people flows, managing access to public parkings, and so onc.
- *White goods:* remote control of washing machines, refrigerators, and so on. In the future, your refrigerator may keep track of its stock, automatically create a shopping list of items missing and order them directly over the Internet.

But M2M communications also introduce new challenges, which explain their slow uptake. Unlike the consumer market, the market for M2M applications is ruled by vertical industries as the previous list shows. M2M applications require the integration of diverse technologies across different business domains, and network operators struggle to create value in industries dominated by established vertical players.

M2M applications will dramatically increase bandwidth demand, which will oblige operators to invest in infrastructure and design more efficient radio networks that employ extremely small cells and take advantage of both licensed and unlicensed spectrum. M2M applications will also accelerate the worldwide deployment of IPv6, which has so far been reluctant.

7.2.2   oneM2M

Network operators see collaboration and standardization as one of the ways to increase their control over the M2M market. This is why leading telecommunications standards organizations from the U.S., Japan, China, Europe, and Korea launched a new partnership project in 2012 called oneM2M (onem2m.org).

Each of oneM2M's member organizations, including ETSI (Europe), ATIS and TIA (U.S.), ARIB (Japan), and CCSA (China), has its own specifications for M2M communications. oneM2M ventures to pull these together into a single model for M2M communications with common services and protocols. Its ultimate mission is to avoid fragmentation and "horizontalize" M2M communications that traditionally belonged to vertical industries using domain-specific protocols and services. oneM2M aims to specify a unified platform that can satisfy the connectivity requirements for a very wide range of M2M devices and applications, from sensors to connected cars and from smart meters to medical equipment.

oneM2M promotes a model with three layers [2], shown in Figure 7.1.

- *Network services layer:* provides basic network services such as 3G and 4G data connectivity, short messaging, and location services;
- *Common services layer:* provides services that are common among vertical M2M applications such as DM, charging, device status and capabilities, and geocoding;
- *Applications layer:* hosts industry-specific M2M applications such as remote metering, blood sugar measurement, and surveillance.

The oneM2M model further distinguishes between the field domain and the infrastructure domain. The field domain is where M2M devices are

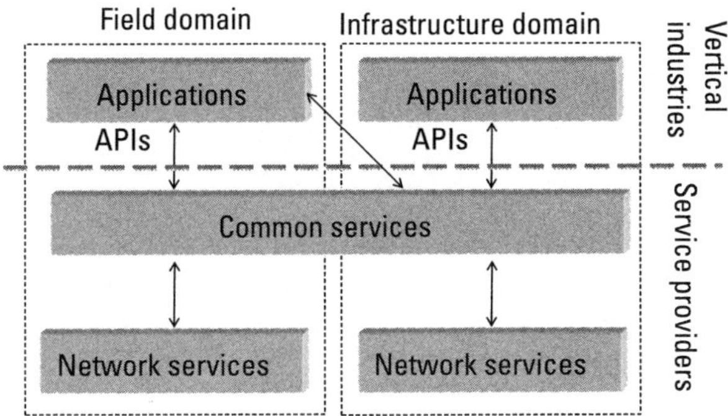

Figure 7.1 oneM2M overall level architecture.

deployed. It typically hosts sensors, actuators, and embedded M2M communications modules and can also include local networks of M2M devices, connected to the infrastructure domain by gateways.

The infrastructure domain covers the computing and networking services that process the data from M2M devices and run the resource-intensive part of M2M applications. The infrastructure domain can include applications that run in the cloud.

7.2.3 oneM2M Common Services Layer

The common services layer represents the critical component of the oneM2M model that provides horizontal functions to M2M applications that belong to vertical industries. With this layer, network operators and service providers intend to provide added value to M2M industries and avoid these vertical markets developing their own, fragmented M2M services. The common services layer includes the following services:

- *Notification*, supporting publish-subscribe type communications between M2M devices and applications (following the Observer design pattern of Section 4.1.5);
- *Device management* (DM), allowing service providers to diagnose and configure M2M devices (see Section 5.3.1 for a brief explanation of DM);
- *Security*, providing common protocols and encryption techniques for resource-constrained M2M devices and data connections;
- *Charging*, allowing M2M applications to charge for the use of services and network resources;

- *Service discovery and brokering*, allowing M2M applications to discover and use common M2M services provided by the operator.

The common services layer provides these services through APIs. In fact, you will have noticed that the common services mentioned here strongly resemble the Parlay X and OMA RESTful network APIs discussed in Sections 5.4.1 and 5.4.2. oneM2M will leverage these existing APIs and reuse them as much as possible. Likewise, oneM2M intends to reuse existing protocols for DM, location, security, and other services.

7.2.4   oneM2M Protocols

oneM2M applications use either RESTful HTTP (see Section 5.2.5) or the constrained application protocol (CoAP) to exchange messages. IETF RFC 7252 defines CoAP [3] as an application-level request-reply protocol that resembles HTTP but optimizes data and computation efficiency for use in resource-constrained devices. More specifically, CoAP has the following characteristics that distinguish it from HTTP:

- CoAP uses a fixed size binary header of only 4 bytes, which is significantly more compact than the plain-text HTTP header.
- CoAP messages must fit in a single IP packet. This restriction does not apply to HTTP and ensures that devices do not have to waste resources by fragmenting CoAP messages over various IP packets and reassembling them at the other end.
- CoAP messages travel as UDP payloads,[1] avoiding the communications overhead that TCP introduces as a result of its need to establish connections and retransmission of lost or corrupted data.
- CoAP merges HTTP style application layer messages GET, POST, PUT, and DELETE with simplified transport layer primitives such as ACK and RESET to manage data flow. In this way CoAP compensates the data transport services that were lost by using UDP instead of TCP, while avoiding the overhead that comes with the latter.

CoAP supports RESTful communications just like HTTP, because it uses similar primitives as HTTP (GET, POST, PUT, DELETE). Putting it simply,

---

[1] CoAP can also use datagram transport layer security (DTLS) if communications need to be encrypted at data transport level. DTLS provides a lean data format to encrypt data transported with UDP.

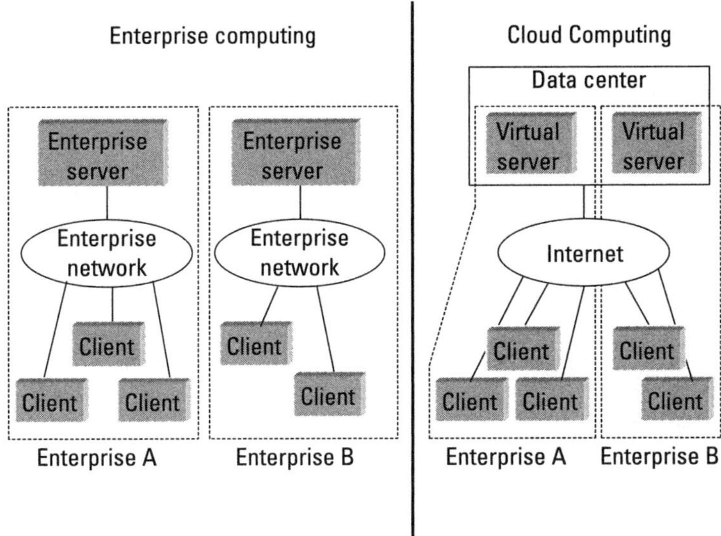

Figure 7.2   Legacy enterprise computing versus cloud computing.

CoAP enables HTTP style communications for devices with constraints on processing power, memory, battery life, and data connectivity.

7.3   Cloud Computing and Virtualization

Cloud computing is one of the technologies that has had the strongest impact on computing technology over the past decade. Cloud computing takes distributed client-server computing to a world scale and can be seen as a logical evolution enabled by the Internet. Some liken the emergence of cloud computing to that of power grids,[2] where large power plants have come to replace local power generation over time.

Figure 7.2 illustrates the evolution from legacy enterprise computing toward cloud computing, where large data centers concentrate computing resources that are available on a pay-per-use basis. These data centers offer economies of scale and can provide elastic computing capacity.

---

[2] Interestingly, power grids now witness a trend featuring local power generation with renewable energy sources and the distribution of grid control to locally managed micro grids. This raises the question of whether we'll see a similar trend in cloud computing at some point. Perhaps cloud computing will eventually merge with grid computing, taking advantage of idle capacity in distributed computing resources that could even include individual PCs and game consoles.

Figure 7.2 does somewhat oversimplify things, however. Although cloud computing and virtualization[3] often go hand in hand, they have different a purpose and they are certainly not synonyms. Cloud computing can exist without virtualization (e.g., when an enterprise installs and runs its software directly on a data center's physical servers and storage systems). Virtualization can also exist without cloud computing (e.g., when an enterprise hosts and manages its own virtualized servers in-house).

7.3.1 Cloud Computing-Enabled Services

Figure 7.2 illustrates the general concept of cloud computing but doesn't show how different types of services can be offered over the cloud. Cloud services are most often divided in three groups:

- *Infrastructure as a service* (IaaS): provides computing resources on which clients can install and run their own operating systems, middleware, and applications. Rackspace® is an example of a company offering raw server infrastructure as a cloud service.
- *Platform as a service* (PaaS): provides computing resources with operating system and middleware on which clients can install run their applications. Amazon Elastic Compute Cloud (EC2)™ is an example of PaaS.
- *Software as a service* (SaaS, sometimes also called service as a service): provides fully managed services and applications that clients can use remotely. Microsoft Office 365™ is an example of SaaS.

Figure 7.3 shows these three types of cloud service in a more graphical way.

7.3.2 Network Operators and Cloud Computing

Network operators can benefit from cloud computing in various ways as users, enablers, and providers. First, they can use third-party cloud computing services to increase the efficiency and scalability of their IT operations just like

---

[3] Virtualization refers to running software on virtual machines, in the way Java byte code executes on a Java virtual machine. Virtual machines (including the Java virtual machine) are enabled by software and managed by APIs, and allow the creation of computing resources on demand. A physical server can run more than one virtual server, and the management software can move jobs between virtual machines to optimize resource usage and load balance. Do keep in mind that a virtual machine translates its code to machine instructions to execute on a physical processor. All code ultimately runs on hardware.

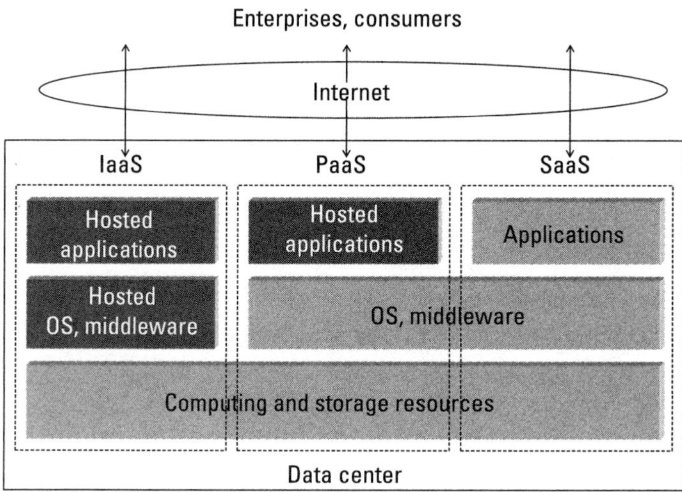

Figure 7.3 IaaS, PaaS, and SaaS.

any other enterprise does. An operator may choose to migrate its own systems to the cloud by contracting IaaS or PaaS from third-party providers. It also becomes more common for network operators to trust part of their operations to third-party-provided SaaS, such as the BSS and IMS signaling or media processing functions.

As enablers, network operators provision network access to third-party data centers that host cloud computing services. Providing networks and connectivity is their core business, after all. But network operators also have a favorable position to become providers of IaaS and PaaS themselves, because they already operate an extensive IT infrastructure with data centers or their own. Figure 7.4 illustrates the roles a network operator can play in the cloud computing ecosystem.

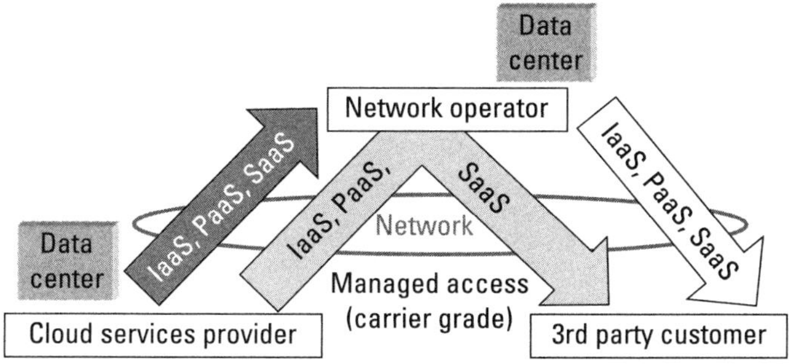

Figure 7.4 The network operator as cloud computing consumer and provider.

Most importantly, network operators own the network that gives access to their data centers, allowing them to manage and dimension network traffic for optimal performance and service. Network operators have a unique opportunity to provide carrier grade cloud services (e.g., music streaming and data storage). They also have unique opportunities to blend fixed and mobile network access, communications services, and cloud-based applications into a compelling user experience.

7.3.3 Communications Services Through the Cloud

An increasingly promising prospect for network operators is to enable communications services through the cloud—in other words, to provide communications SaaS. Looking back at Chapter 3, you'll realize that IMS has an all-IP computing infrastructure. Even legacy systems such as IN and CAMEL that use SS7 signaling essentially consist of computing nodes connected by a data network. Network operators can offer cloud-based processing of IMS signaling to third-party service providers without their own network infrastructure. Cloud-based servers can also host the network APIs, as described in Chapter 5. This allows network operators to let MVNOs exploit their wholesale communications services.

The future is likely to see more complex ecosystems whereby network operators and other SaaS providers provide cloud-based services to each other to create new business engagements. We can particularly expect this in the M2M domain where vertical industries depend on highly specialized, networked applications. Figure 7.5 illustrates such new types of relationships between network operators and cloud services providers form vertical industries.

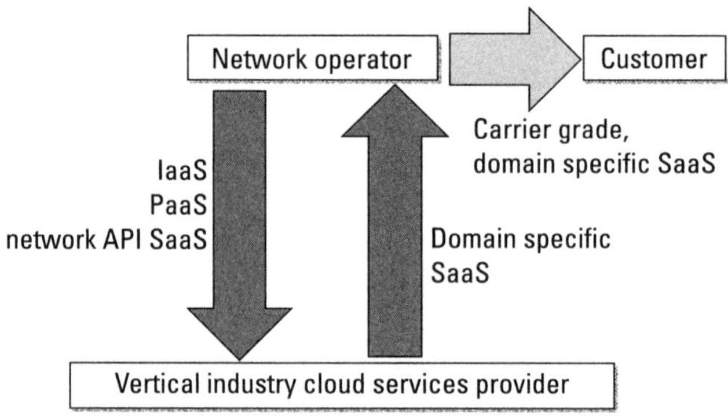

Figure 7.5 New cloud service business models for network operators.

7.3.4 Cloud Computing Challenges

Although cloud computing brings clear opportunities, not all is bliss. As cloud computing becomes a mainstream technology, new challenges also appear. Some of the technical issues that the cloud community has to address in the near future are as follows:

- *Security:* despite the use of state-of-the-art identification, authentication, authorization, and encryption technologies, news about high-profile security breaches emerges almost daily. The problem does not lie in the security algorithms themselves, which have mathematically proven integrity, but in how they are programmed. In practice it's almost impossible to avoid bugs in enterprise scale programs that can comprise millions of lines of code. A hacker can exploit even the smallest oversight or bug, which can render security protocols useless, no matter their theoretical strength. Hackers also exploit the human factor, which most often represents the weakest link in a system. Many security breaches begin with social engineering[4] or phishing.

- *Traffic management:* as cloud computing advances and data centers grow in numbers and size, they concentrate enormous amounts of data traffic. According to an article in online technology magazine *Wired*, Amazon concentrates 1% of the all Internet's traffic. Think of what that means: one out of every hundred IP packets on the Internet goes through Amazon's data centers. Network connections to large data centers require careful dimensioning and planning, and we often overlook the fact that enterprises using cloud services also need increasingly high bandwidth connections.

  Virtualization introduces additional challenges to bandwidth planning and management. Virtualization tends to make traffic less predictable, as computing jobs can move between virtual machines for optimized resource use and load balancing.

---

[4] Social engineering consists of convincing or tricking a human subject into giving up personal and classified information, typically a username-password combination for an online account or system. Social engineering may involve befriending a company's staff or randomly calling a company's employees and trying to make them give up personal account information (e.g., by impersonating the company's IT department). Phishing is a specific form of social engineering whereby a hacker sends e-mail or chat (usually in bulk to many users) and impersonates a party that the user trusts to obtain classified information.

- *Resource use.* Data centers use enormous amounts of electricity—about 1.5% of the world's total electricity consumption. This means that data centers bear responsibility for the emission of significant amounts of carbon dioxide (CO_2), the gas that causes the greenhouse effect linked to global warming. Energy efficiency has become a priority, not only to keep operational costs of datacenters down but also to reduce their environmental impact.

Much of the electricity used by data centers goes into the systems that cool the computing equipment. The European Commission and other organizations have been funding research into reusing the heat generated by data centers for heating homes or public buildings, or even generating electricity to be reinjected into the power grid.

These are considerable challenges that need innovative solutions. A great deal of research presently goes into making cloud computing more secure and energy efficient, and making sure that the Internet can keep up with the new traffic patterns it creates.

Cloud computing also introduces more business-related challenges such as compatibility between different cloud computing technologies, standards fragmentation, and cloud service providers' strategies to lock customers in and make it difficult for them to migrate to competitors. This means that cloud computing will remain a dynamic area where new opportunities and threats emerge as technology advances.

7.4   Network Centric Versus Over the Top Services

From the invention of telephony in the late nineteenth century until the 1980s, telecommunications hardly underwent any significant changes. Though manual exchanges made way for electromechanical and later digital exchanges, the telephone and telex service essentially remained the same. The telephone operator's network provided the services and terminals were passive devices without processing capacity. Even in the era of advanced computing technologies, IN and CAMEL still adhered to this model, as we saw in Chapter 2.

The Internet started uprooting this communications model in the 1990s. Conceived as a data network, the Internet connects devices that run applications and store data. On the Internet, computing devices do the processing and the network just transports data. Many applications on the Internet use peer-to-peer communications without central control. In applications that use a client-server model, the server is just another host[5] connected to the Internet and does not form part of the Internet itself.

7.4.1 The Smart Phone Revolution

Mobile networks have undergone the same transition from network- to device-based control. The 2G networks such as GSM follow the telephony model, where the network controls the service and the mobile device is essentially a telephone. The 3G networks brought the Internet to mobile devices, but initially these devices offered only basic Internet applications such as web browsing and e-mail.

The emergence of smart phones and tablets around 2007 marked a true turning point. Smart phones represent a crossover between a telephone and a lightweight computer with high-definition graphics and sound that can run more sophisticated applications. This proved to be a sweet spot in the consumer market: a pocket device that serves as mobile phone, camera, media player, web browser, and game console.

Mobile network operators saw the high price tag of the early smart phones as an opportunity for subsidizing them as a way to lock in customers. Subsidizing smart phones has been instrumental in their explosive growth and has allowed them to mature rapidly.

Smart phones went through a similar evolution as personal computers and are now commoditized to price points that make operator subsidies pointless. On the other hand, the proliferation of smart phones has fiercely increased the amount of data traffic, obliging mobile network operators to invest heavily in infrastructure to keep up with the growing demand for data communications. By subsidizing smart phones, mobile network operators accelerated a revolution that is now challenging their very business model.

Not only smart phone technology brought about these changes. Online application stores such as Apple's App Store or Google's Play Store gave the decisive boost to the creation of an active ecosystem of application producers and consumers. Application stores provide two important advantages over having developers publish their applications directly on the Internet.

First, they provide a one-stop shop for application consumers and a distribution channel for application developers. Application stores handle

---

[5] In IETF terminology, a host is a device with an IP address connected to the Internet. Technically speaking, a host can be anything from a firewall to a server or PC, tablet, or smart phone. A host can have more than one IP address, in which case it's called *multihomed*.

payment and download, and guarantee that the developer gets paid and the consumer gets their product. In turn, it takes a share of the revenue. Second, application stores review applications before publishing them. This gives the consumer confidence that the application works and does not contain malware.

7.4.2 The Operators' Answer: IMS and WAC

Smart phones and application stores have profoundly changed the face of telecommunications, drawing control away from the network and into the device. Figure 7.6 illustrates this shift from control in the network toward control by third-party application servers.

Network operators responded to this trend with IMS (Chapter 3) and by creating their own application store architecture called the wholesale application community (WAC). Whereas Google's Play Store and Apple's App Store provide applications for a particular device operating system, WAC was designed to let network operators draw from a general repository of applications for a variety of operating systems to customize an application offering for their particular market. Where iOS and Android applications can have an important amount of native code running on the device, WAC applications rely almost exclusively on cloud-based computing. This makes it easier to provide the same application on different operating systems and gives operators more control over the application and its connectivity.

WAC headed for failure almost from its inception, and GSMA stopped supporting WAC only two years after its launch in 2010. Analysts gave several reasons for its failure, but Apple's and Google's tenacious grip on the applications market probably played an important role.

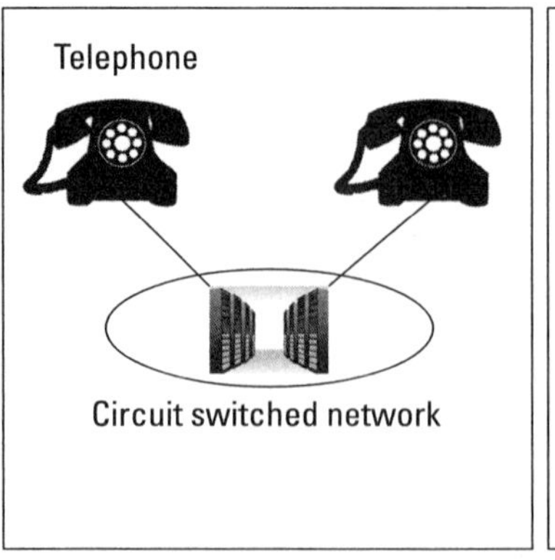

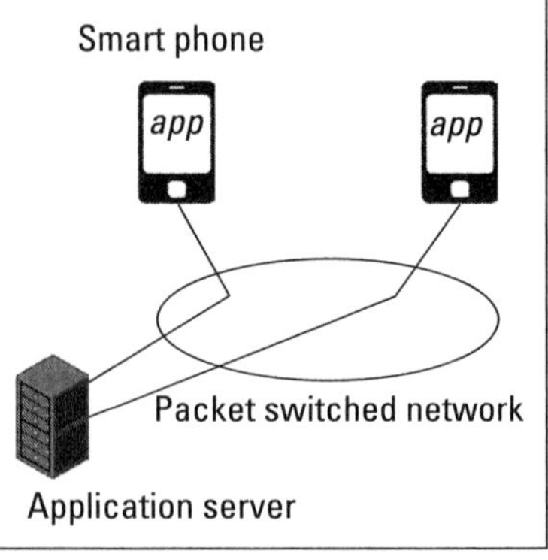

Figure 7.6 Control inside and outside the network.

Throughout its first decade in existence, IMS has been struggling to become widely deployed by network operators. But the need for a standardized architecture to enable voice telephony over IP in LTE networks resuscitates interest in IMS.

This leaves us with intriguing questions for the future: will the network or the device control value-added applications and services? Who will control the communications services of the future: the network operator or third-party OTT providers? These questions challenge the very business of the network operator, and we'll discuss them in the last section of this book.

7.5 The Future of the Network Operator

This book dedicates most of its chapters to technologies that network operators deploy to support their business, from IN and CAMEL to IMS, Java and network APIs to OSS and BSS. But as we've seen in this chapter, recent trends have put pressure on the network operators' business models.

7.5.1 Bumps Ahead

Network operators have their hopes on LTE as an enabler for faster and superior quality data services that will make revenues grow again. But some analysts warn that network operators could accelerate their own demise with LTE. They argue that revenues from faster data services will fail to offset the increase in operational cost. What's worse, by providing faster and better data services, network operators only help third parties increase the market for OTT applications.

Subsidizing devices, once considered a good strategy to capture and retain subscribers, can also backfire on network operators. By subsidizing devices, network operators have commoditized them and have contributed to the surge in demand for data traffic that has been hard and expensive to keep up with.

Operators hope that M2M applications can bring growth beyond the saturated market for personal communications, but here again the question of where the real value lies remains: in the data enablers or in the domain-specific applications provided by third parties from vertical industries?

7.5.2 After Rain Comes Sunshine

But it's not all malaise for network operators, as they have many strategic assets including radio spectrum, network infrastructure, carrier-grade services, subscriber data analytics, and a billing framework. Each of these provides opportunities for new business models. In the coming years, operators will face the challenge to find the pot of gold at the end of the rainbow.

A. T. Kearney [4] sees opportunities for network operators in three areas:

- *APIs:* providing interfaces to (payable) network functions that can be used by third parties to add value to their applications.
- *Ecosystem:* setting standards and forging agreements across sectors and industries. As they provide network access and have a trusted relationship with their subscribers, network operators are in an excellent position to provide value-added services and applications by brokering collaboration between different players.
- *Customer data analytics:* operators have privileged access not only to subscriber data, but also to information on how (and where) their subscribers use the services provided by the network. These analytics have value for third parties including providers of OTT applications.

Despite the competition from OTT applications, network operators have an opportunity to control the user experience through their network and subsidized devices. Unlike any other player, network operators can integrate fixed and mobile telephony, data, and broadcast services to provide a seamless communications experience and build a compelling service offer that OTT players have more problems providing. This does pit them directly against big players such as Google and Apple, who also build their business on providing a complete user experience. Operators have so far avoided this direct confrontation and have rather sought to differentiate themselves from these IT players by emphasizing their role as telecommunications provider. Sooner or later they may have to challenge OTT players head on.

Network operators also have an edge over OTT players in the enterprise communications domain. While the enterprise market has a lower volume than the consumer market, enterprises still depend on network operators for carrier-grade communications that OTT players have difficulty providing.

Meanwhile, operators explore new ways of combining competition and collaboration to reduce operating cost. Network operators increasingly turn to sharing assets as a way to reduce cost and increase coverage and efficiency. They can share passive assets such as towers, cell sites, and real estate or active assets such as base stations or even entire networks. Asset sharing becomes an option when a network operator can no longer differentiate and make sustainable revenue just by owning an access network and providing coverage for voice and data services.

7.5.3 Regulation to the Rescue

Although regulation has upset rather than benefitted operators' business by increasing competition and demanding prices come down, regulation can ultimately defend the network operator's business as well. Future regulation could allow network operators to charge OTT players for delivering applications to customers over their networks. Although this challenges network neutrality,[6] one could argue that also it has certain logic to it.

In another scenario, regulation would treat fixed and mobile networks as public infrastructures just like public roads, railroads, and power grids. The regulator could adjust the degree of nationalization and privatization to guarantee a level of service that is in public interest, regardless of whether that service is profitable. This could even lead to the nationalization of some of the network operators' infrastructure, which would mark a return to the days when in some countries, network operators were state-owned companies.

7.5.4 Que Sera, Sera

Network operators face many challenges, yet have many key assets and many possibilities to create business with them in a shifting market. The game has changed, but operators still own the game board. The one thing operators cannot afford today is complacency and clinging to legacy business models. This is understandably difficult for large corporations, some of which have enjoyed steady business from telephony and telegraphy for almost a century.

The future will be difficult to predict, but, no matter what happens, it will be interesting and it will bring fresh surprises. Some of the technologies discussed in this book will enjoy success, others will not. New disruptive technologies may appear. What is certain is that we'll continue to communicate in new and perhaps unexpected ways.

References

[1]   Zuidweg, J., *Next Generation Intelligent Networks*, Norwood, MA: Artech House, 2002.

[2]   oneM2M, oneM2M Functional Architecture Baseline, oneM2M-TS-0001, 2014.

---

[6] The network neutrality principle requires that Internet service providers must treat all data traffic equally and shall not discriminate IP packets on the basis of their origin, destination, content, or whatever other criteria. Network neutrality is a controversial and highly debated subject. Some countries regulate network neutrality by law; others do not.

[3]   Shelby, Z., K. Hartke, and C. Bormann, The Constrained Application Protocol (CoAP), RFC 7252, IETF, 2014.

[4]   Kearney, A. T., Mobile Network Operators as "Smart Enablers," white paper, 2012.

[5]   3GPP, Digital Cellular Telecommunications System (Phase 2+), Universal Mobile Telecommunications System (UMTS), Mobile Execution Environment (MExE), Functional Description, Stage 2, TS 23.057, 2012.

Glossary

2.5G   Data-enhanced second generation digital cellular networks

2G   Second generation ditigal cellular networks

3G   Third generation digital cellular networks

3GPP   Third Generation Partnership Project

3GPP2   Third Generation Partnership Project 2 (parallel project)

4G   Fourth generation digital cellular networks

AAA   Authentication, authorization, and accounting

ABMF   Account balance management function

ADSL   Asymmetric digital subscriber line

AIN   Advanced intelligent network

ANSI   American National Standards Institute

AOC   Advice of charge

API   Application programming interface

ARIB   Association of Radio Industries and Businesses (Japan)

ARP   Address resolution protocol

ARP   Alternative roaming provider

AS   Application server

AS-ILCM   Application server incoming leg control model

ASN.1   Abstract syntax notation no. 1

AS-OLCM   Application server outgoing leg control model

ATM   Asynchronous transfer mode

BCP   Basic call process

BCSM   Basic call state model

BGCF   Breakout gateway control function

BGP   Border gateway protocol

BGW   Border gateway

BS   Base station

BSC   Base station controller

BSS   Base station subsystem

BSS   Business support system

BT   British Telecom

BTS   Base transceiver system

CAMEL   Customized applications for mobile network enhanced logic

CAP   CAMEL sscol

CB   Call barring

CC/PP   Composite capability/preference profile

CCBS   Completion of communications to busy subscriber

CCF   Connection control function

CCNR   Completion of communications on no reply

CCSA   China Communications Standards Association (P.R. China)

CDC   Connected device configuration

CDF   Charging data function

CDMA   Code division multiple access

CDR   Call detail record

CGF   Charging gateway function

CID   Call instance data

CLDC   Connected limited device configuration

CMIP   Common management information protocol

CoAP   Constrained application protocol

COBOL   Common business oriented language

CORBA Common object request broker architecture

CPH Call party handling

CPL Call processing language

CPM Converged IP messaging

CRACF Call-related radio access control function

CRM Customer relationship management

CRS Customize ringing signal

CS Call segment

CS Capability set

CS-1 Capability set 1

CS-2 Capability set 2

CS-3 Capability set 3

CS-4 Capability set 4

CSA Call Segment Association

CSCF Call state control function

CSI CAMEL subscription information

CTF Charging trigger function

CURACF Call unrelated radio access control function

CUSF Call unrelated service function

CVS Connection view state

CW Call waiting

DAMPS Digital AMPS

DAP Directory access protocol

D-CSI Dialed services CSI

DM Device management

DNS Domain name system

DP Detection point

DSO Standards Development Organization

DSP Domestic service provider

DTD Document type definition

DTMF Dual tone multifrequency

E.164 International numbering plan standard (ITU)

EBCF   Event-based charging function

EC2   Elastic Compute Cloud (Amazon trademark)

EDGE   Enhanced data rate for GSM evolution

EDP   Event detection point

EDP-N   Event detection point (notification type)

EDP-R   Event detection point (request type)

EJB   Enterprise Java beans

eTOM   Enhanced telecommunications operations map

ETSI   European Telecommunications Standards Institute

FCAPS   Fault, configuration, accounting, performance, and security

FE   Functional entity

FTP   File transfer protocol

G.711   PCM codec standard (ITU)

G.726   ADPCM codec standard (ITU)

G.728   LD-CELP codec standard (ITU)

G.729   CS-ACELP codec standard (ITU)

GCF   Global Certification Forum

GERAN   GSM-EDGE radio access network

GFP   Global functional plane

GGSN   Gateway GPRS support node

GMSC   Gateway mobile switching center

GNSS   Global navegation satellite system

GPRS   General packet radio system

GPS   Global positioning system

GSM   Global System for Mobile Communications

GSMA   GSM Association

H.225   Remote access server standard (ITU)

H.245   Control signaling standard (ITU)

H.261   Video codec standard (ITU)

H.263   Video codec standard (ITU)

H.320   Multimedia over ISDN standard (ITU)

H.323   Multimedia over packet networks standard (ITU)

HLR Home location register

HLSIB High-level SIB

HSS Home subscriber server

HTML Hyper text markup language

HTTP Hyper text transport protocol

IaaS Infrastructure as a service

IAB Internet Architecture Board

IAF Intelligent access function

IANA Internet Assigned Numbers Authority

IARI IMS application reference identifiers

IBCF Interconnection border control function

ICID IMS charging identifier

ICMP Internet control message protocol

I-CSCF Interrogating call state control function

ICSI IMS communication service identifiers

IDL Interface definition language

IESG Internet Engineering Steering Group

IETF Internet Engineering Task Force

IFC Initial filter criteria

IGMP Internet group management protocol

IIOP Internet inter ORB protocol

ILCM Incoming leg control model

ILSM Incoming leg state model

IMP Information module profile

IMS Internet multimedia subsystem

IMSI International mobile subscriber identity

IMT-2000 International mobile telecommunications

IMT-A International mobile telecommunications advanced

IN Intelligent network

INAP IN application protocol

INCM Intelligent network conceptual model

IOI Interoperator identifier

IoT   Internet of things

IOT   Interoperability testing

IP   Internet protocol

IP   Intelligent peripheral

IPDR   IP detail record

IPv4   Internet protocol version 4

IPv6   Internet protocol version 6

IPx   IP interconnect

ISDN   Integrated services digital network

ISDN   Integrated services digital network

ISO   International Standards Organization

ISOC   Internet Society

ISP   Internet service provider

ISUP   ISDN user part

ITU   International Telecommunications Union

JAIN   Java APIs for integrated networks

JAIN SLEE   JAIN service logic execution environment

JCAT   Java coordination and transactions

JCC   Java call control

JCP   Java call processing

JEE   Java enterprise edition

JME   Java micro edition

JSE   Java standard edition

JSON   JavaScript object notation

JSR   Java specification request

KPI   Key performance indicator

KQI   Key quality indicator

LAI   Location area identifier

LDAP   Lightweight directory access protocol

LISP   List processing (programming language)

LTE   Long-term evolution

LTE-A   Long-term evolution advanced

M2M   Machine to machine

MAP   Mobile application part

MExE   Mobile execution environment

MGCF   Media gateway controller function

MGW   Media gateway

MIB   Managemen information base

MIDP   Mobile information device profile

MIME   Multipurpose internet mail extensions

MME   Mobility management entity

MMTel   Multimedia telephony

MO   Managed object

MPEG   Moving Picture Experts Group

MS   Mobile station

MSC   Mobile switching center

MS-ISDN   Mobile station ISDN number

MSRN   Mobile subscriber roaming number

MTBF   Mean time between failures

MTOSI   Multitechnology operations system interface

MTP   Message TRANSFER PArt

MTTR   Main time to recovery

MVNO   Mobile virtual network operator

NAT   Network address translation

N-CSI   Network-specific services CSI

NSS   Network switching subsystem

OAM   Operations, administration, and management

O-BCSM   Originating basic call state model

OCS   Originating call screening

O-CSI   Originating call CSI

Ofcom   Office of Communications (United Kingdom)

OLCM   Outgoing leg control model

OLSM   Outgoing leg state model

OMG   Object Management Group

oneM2M   Machine-to-Machine Partnership Project

OpenCMAPI   Open connection management API

ORB   Object request broker

OSA   Open service access

OSI   Open systems interconnection

OSPF   Open shortest path first

OSS   Operation support system

OSS/J   OSS through Java

OTT   Over the top

PaaS   Platform as a service

PBX   Private branch exchange

PC   Personal computer

P-CSCF   Proxy call state control function

PDF   Portable document format

PDN-GW   Packet data network gateway

PDP   Packet data protocol

PIC   Point in call

PICS   Protocol implementation conformance statement

PIDF   Presence information data format

PIN   Personal identification number

PINT   PSTN and Internet Interworking Working Group

POI   Point of initiation

POP3   Post office protocol (version 3)

POR   Point of return

POS   Points of sale

PP   Physical plane

PROLOG   Programming with logic (programming language)

PSTN   Public switched telephone network

PUK   Personal unblocking code

QoS   Quality of service

RARP   Reverse address resolution protocol

RCF   Radio control function

RCS   Rich communications services

REST   Representational state transfer

RFC   Request for comments (IETF)

RMI   Remote method invocation

RPC   Remote procedure call

RTCP   Real-time transport control protocol

RTP   Real-time transport protocol

SaaS   Service as a service

SAT   SIM application toolkit

SBB   Service building block

SBCF   Session-based charging function

SCCP   Signaling connection control part

SCE   Service creation environment

SCF   Service control function (IN), service capability feature (OSA)

SCIM   Service capability interaction manager

SCP   Service control point

S-CSCF   Serving call state control function

SCUAF   Service control user agent function

SDF   Service data function

SDK   Software development kit

SDL   Specification and description language

SDP   Session description protocol

SDP   Service data point

SDP   Service delivery platform

S-GN   Serving gateway

SGSN   Serving GPRS support node

SIB   Service independent building block

SID   Shared information and data

SIM   Subscriber identity module

SIMPLE   SIP for instant messaging and presence leveraging extensions

SIP   Session initiation protocol

SLA   Service level agreement

SMAF   Service management access function

SMF   Service management function

SMP   Service management point

SMS   Short message service

SMTP   Simple mail transfer protocol

SNMP   Simple network management protocol

SOAP   Simple object access protocol

SP   Signaling point (in SS7), or service plane (in IN conceptual model)

SPIRITS   Services in the PSTN/IN requesting Internet services (IETF)

SQL   Structured query language

SRF   Specialized resource function

SS7   Signaling system No.7

SSD   Service support data

SSF   Service switching function

SSL   Secure socket layer

SSP   Service switching point

STBP   Set-top box profile

STP   Signaling transfer point

T-BCSM   Terminating basic call state model

TCAP   Transaction capability application part

TCP   Transport control protocol

TCP/IP   TCP over IP

TCS   Terminating call screening

T-CSI   Terminating call CSI

TDP   Trigger detection point

TDP-N   Trigger detection point (notification type)

TDP-R   Trigger detection point (request type)

TIF   Translation information flag

TMN   Telecommunications management network

TMSI   Temporal mobile subscriber identity

TrGW   Transition gateway

TTA   Telecommunications Technology Association (Korea)

TTC Telecommunication Technology Committee (Japan)

TTL Time to live

TUP Telephone user part

UA User agent

UAC User agent client

UAProf User agent profile

UAS User agent server

UDP User datagram protocol

UDP/IP UDP over IP

UML Unified modeling language

UMTS Universal mobile telecommunications system

USAT USIM application toolkit

USSD Unstructured service support data

UTRAN UMTS terrestrial radio access network

VDSL Very high speed digital subscriber line

VLR Visitor location register

VoIP Voice over IP

VPN Virtual private network

VT-CSI Visited network terminating call CSI

W3C World Wide Web Consortium

WAC Wholesale applications community

WAP Wireless application protocol

WDP Wireless datagram protocol

WiFi Wireless local area networks based on the Institute of Electrical and Electronics Engineers' 802.11 standards

WIN Wireless intelligent network

WML Wireless markup language

WS Web services

WSDL Web services definition language

WSP Wireless session protocol

WWW World Wide Web

X.25 Packet data networking standard (ITU)

X.500   Directory service standard (ITU)

XCAP   XML Configuration access protocol

XDM   XML document management

XDR   External data representation

About the Author

Johan Zuidweg has an ICT consultancy called Next Generation Networks Research (www.ngnresearch.com) and incubates Internet startups. He has also been a lecturer at the Pompeu Fabra University in Barcelona, Spain, since 2001 and has received several distinctions for his teaching of computer science and telecommunications courses.

Over the past 25 years, Johan has been working for network operators (Royal Dutch KPN and Telefónica), telecommunications equipment vendors (Alcatel-Lucent and Huawei Technologies), and a system integrator (Tecsidel), which enabled him to experience communications technologies and markets from different perspectives. Throughout his carreer, he has been at the forefront of technological innovation. In the early 1990s he coordinated the specification of UMTS (the European 3G mobile network standard) as a member of the R1043 UMTS Systems Group in Paris. In the second half of the 1990s, he helped design the world's first distributed object-oriented IN system at Alcatel in Belgium and lead the European project VITAL, which successfully built and tested an innovative telecommunications architecture based on distributed object-oriented computing principles.

Johan presently chairs the ETSI industry specification group (ISG) for the specification of a machine-to-machine communications module based on surface mount technology (SMT) and has been leading groups in the Open Mobile Alliance (OMA) between 2006 and 2014.

Johan has previously published three books on next generation networks and services, two of which are with Artech House. He has authored numerous scientic articles and holds a number of U.S., European, and Chinese patents in the area of identity management and content filtering.

Johan holds a Ph.D. and M.Sc. in Computer Science from Leiden University in the Netherlands. He is a member of the IEEE Computer and Communications Societies.

You can contact Johan at johan.zuidweg@ieee.org or johan.zuidweg@ gmail.com.

Index

Understanding Voice over IP Security, Alan B. Johnston and
David M. Piscitello

*Videoconferencing and Videotelephony: Technology and Standards,
Second Edition*, Richard Schaphorst

Visual Telephony, Edward A. Daly and Kathleen J. Hansell

Wide-Area Data Network Performance Engineering, Robert G. Cole and
Ravi Ramaswamy

Winning Telco Customers Using Marketing Databases, Rob Mattison

WLANs and WPANs towards 4G Wireless, Ramjee Prasad and Luis Muñoz

World-Class Telecommunications Service Development, Ellen P. Ward

For further information on these and other Artech House titles, including previously
considered out-of-print books now available through our In-Print-Forever® (IPF®) program, contact:

Artech House
685 Canton Street
Norwood, MA 02062
Phone: 781-769-9750
Fax: 781-769-6334
e-mail: artech@artechhouse.com

Artech House
16 Sussex Street
London SW1V HRW UK
Phone: +44 (0)20 7596-8750
Fax: +44 (0)20 7630-0166
e-mail: artech-uk@artechhouse.com

Find us on the World Wide Web at: www.artechhouse.com